READy
Renewable Energy
Action on Deployment

READy

RENEWABLE ENERGY

ACTION ON DEPLOYMENT

Presenting: The ACTION Star; six
policy ingredients for accelerated
deployment of renewable energy

IEA-RETD

ROLF DE VOS
Ecofys Netherlands BV, Utrecht, The Netherlands

JANET SAWIN
*Sunna Research
LLC, Sandwich
New Hampshire*

AMSTERDAM • BOSTON • HEIDELBERG • LONDON
NEW YORK • OXFORD • PARIS • SAN DIEGO
SAN FRANCISCO • SYDNEY • TOKYO

ELSEVIER

Elsevier
The Boulevard, Langford Lane, Kidlington, Oxford OX5 1GB, UK
Radarweg 29, PO Box 211, 1000 AE Amsterdam, The Netherlands
225 Wyman Street, Waltham, MA 02451, USA

First edition 2012

Library of Congress Cataloging-in-Publication Data
Application submitted

British Library Cataloguing in Publication Data
A catalogue record for this book is available from the British Library

ISBN: 978-0-12-405519-3

For information on all Elsevier publications visit
our website at store.elsevier.com

Printed and bound by CPI Group (UK) Ltd, Croydon, CR0 4YY

This READy book has been authored and edited by Rolf de Vos (Ecofys) and Janet Sawin (Sunna Research), on behalf of the IEA-RETD. The authors thank Kristin Seyboth (Seyboth and associates) for her text contributions concerning renewable heating and cooling. The authors were also assisted and advised by Eric Martinot, by Ecofys consultants and by members of the IEA-RETD Executive Committee in finding the right data, the inspiring case studies and the adequate tone of voice in the book.

CONTENTS

FOREWORD

Accelerating renewable energy technology deployment

Renewable energy technologies are on the verge of a new era. In many countries and regions, renewable energy is already responsible for meeting a substantial share of energy demand. The rapid and substantial progress of renewable energy in recent years has been driven by policies of local, national and regional authorities, in close cooperation with the business community, as well as continued technological innovation and cost reductions in energy generated with renewable sources.

Progress in building new energy systems is already considerable. But issues like energy independence, eradicating energy poverty, combating climate change and improving the crisis-robustness of energy systems are asking to accelerate the deployment of renewables. Recent events that have had large impacts on societies around the world—e.g., the financial crisis, the nuclear accident in Fukushima, large oil spills, new findings in climate change science—have further highlighted this need.

However, there are also some trends that mask the opportunity of accelerating renewable energy growth. For instance, the promising—short-term—benefits from exploring new fossil sources like oil in arctic regions or shale gas are mobilizing powerful forces that tend to disregard the long-term disadvantages. Meanwhile, in debates the real costs of renewables are often misrepresented or misperceived, and the reality of integrating decentralized renewable sources in grids is often misperceived as a hurdle that cannot be taken.

One should acknowledge the *real* challenges of large-scale deployment of renewable technologies, which requires both institutional, technological and societal change. But the challenges of a business-as-usual strategy outweigh a renewable route by orders of magnitude: dealing with increased and volatile oil prices, insecurity of energy supply, climate change, air pollution, major accidents, et cetera.

So here is nowadays' challenge for policy makers and decision makers: how to pass the threshold in the short term in order to prepare for the longer term? Achieving energy systems that will meet tomorrow's energy demand in a sustainable, responsible way is possible, as some countries already prove. But further deployment requires a large effort on the part of policy makers and business leaders.

The International Energy Agency's Implementing Agreement on Renewable Energy Technology Deployment (IEA-RETD) is a group of nine countries that advocate a significantly higher utilization of renewable energy technologies. RETD believes international cooperation and public-private partnerships are crucial means to establish a more rapid and efficient deployment, and that it is important to step up to the plate today.

This READy book has been commissioned for the purpose of inspiring and guiding action to begin now. To the readers—policy makers and other decision makers—the READy book presents a variety of policy options that have proven to accelerate the deployment of renewable energy technologies, based on experiences around the world at the local to national to regional levels. Lessons learned from successful cases reviewed in the book are distilled into six essential action points. Together these categories of policy actions compose the ACTION Star, a guide for taking action now while preparing for growth over the long term.

Decision and policy makers will find inspiration in the significant renewable energy developments to date, the many examples of successful policies in this book, and the ACTION star which provides policy recommendations for the way forward in six categories of actions that policy makers can begin to take now.

Hans Jørgen Koch
Chairman of the International Energy Agency's Implementing Agreement on Renewable Energy Technology Deployment (IEA-RETD)
Denmark Deputy State Secretary of Energy

ABOUT IEA-RETD

Renewable energy will play a critical role in de-carbonizing the energy sector, reducing the costs associated with climate change impacts and adaptation, providing energy access to all, and securing long-term energy security at affordable costs. However, it is widely recognized that the establishment of a more sustainable, low-carbon energy system, based largely on renewable energy, will require a major transition in the energy sector and beyond in both scale and scope.

The International Conference for Renewable Energies in Bonn, Germany, in June 2004, was the first of a series of major international conferences to address questions such as how to substantially increase the share of renewables in global energy supply, and how to better make use of their potential and advantages. Participants—including government ministers, representatives of the United Nations and other international and nongovernmental organizations, civil society, and the private sector—aimed to chart the way toward accelerated deployment of renewable energy.

The IEA-RETD Implementing Agreement was one of the key outcomes of the Bonn conference. This knowledge exchange framework was established to focus on how to bridge the gap, generally called the "Valley of Death," between research and development (R&D) and deployment of renewable energy technologies. In Bonn and at relevant follow-up conferences, countries at different stages of development all over the world concluded that benefiting from each other's experiences and best practices will be crucial, and thus they came together to work within the RETD framework.

The IEA-RETD is currently comprised of nine countries: Canada, Denmark, France, Germany, Ireland, Japan, the Netherlands, Norway, and the United Kingdom. The IEA-RETD is a policy-focused, technology cross-cutting platform that brings together the experience and best practices of countries, along with the expertise of renowned consulting firms and academic institutions. IEA-RETD believes that a stable and predictable policy framework is required to create the conditions in which renewable energy can be deployed swiftly and economically at the scale required to address major energy related challenges.

The IEA-RETD is one of a number of Implementing Agreements on renewable energy under the framework of the International Energy Agency (IEA).

IEA-RETD OBJECTIVES AND TARGET GROUPS

Already, the renewable energy industries have made enormous progress and taken substantial steps along the learning curve, and many renewable energy technologies are experiencing high rates of market growth. Nevertheless, policy and decision makers require a step change to stimulate wider deployment of renewable energy technologies and to avoid further lock-in effects of investments in conventional energy technologies.

IEA-RETD aims to demonstrate the need for action and to motivate relevant players to take advantage of the current window of opportunity. The IEA-RETD framework intends to empower energy policy makers and energy market actors through the provision of information and tools, and to provide possible pathways toward accelerated deployment and commercialization of renewable energy. Some general IEA-RETD objectives include the following:

- Make transparent and demonstrate the impact of renewable energy action and inaction
- Facilitate and show best practice measures
- Provide solutions for leveling the playing field between renewable energy and other energy resources and technologies
- Make transparent the market frameworks for renewable energy, including infrastructure and cross-border trade
- Demonstrate the benefits of involving private and public stakeholders in the accelerated deployment of renewable energy technologies
- Enhance stakeholder dialog
- Implement effective communication
- Organize outreach activities

DISCLAIMER

This book draws heavily on studies commissioned by the IEA-RETD as well as reports produced by other international and national organizations. Its content does not necessarily reflect the views and opinions of the IEA Secretariat or of its individual member countries.

READy: Renewable Energy Action on Deployment
Executive Summary

The ACTION star

The ACTION star is a graphic representation of six key policy actions that the RETD recommends for successful acceleration of renewable energy deployment. It is a simple tool for policy makers to develop and analyze their portfolio of policies for renewable energy deployment.

ACTION is the acronym for six required categories of policy actions:

 A: Build alliances and reach agreements among policy makers and with relevant stakeholders including industry members, consumers, investors, and others.

 C: Communicate and gather knowledge about renewable energy resources, technologies, and issues to create awareness on all levels, address concerns of stakeholders, and build up the needed work force.

 T: Clarify the goals and set ambitious targets on all levels of government, and enact policies to achieve goals.

 I: Integrate renewables into policy making related to social, economic, and technical structures and systems, and take advantage of synergies with energy efficiency.

 O: Optimize by building on own policies or other proven policy mechanisms and adapting them to specific circumstances.

 N: Neutralize disadvantages in the marketplace, such as misconceptions of costs and the lack of a level playing field.

SIX POLICY ACTIONS FOR ACCELERATED DEPLOYMENT

The **READy** book presents a kaleidoscope of policy options that have proven to accelerate the deployment of renewable energy technologies, based on experiences around the world at the local to national levels. Lessons learned from successful cases reviewed in the book are distilled into six essential action points. Together these categories of policy actions compose the ACTION Star, a guide for taking action now while preparing for growth over the long term.

> Policy makers play a key role in accelerating renewable energy deployment.

There is growing consensus that a transformation of the energy system must begin immediately. This is because investment decisions made today could lock countries onto a particular path for the next several decades, and because any delay will increase the economic costs associated with energy production and use as well as the costs of the required transition. Thus, this publication focuses on actions that are needed now.

Working in close cooperation, policy makers and the business community can bring about necessary and timely changes in the energy system. This collaboration enables a smooth transition to an economy that is based primarily on renewable energy sources. Policy makers play a key role in accelerating deployment of renewable energy technologies by influencing near- and long-term planning and investment decisions through government policy.

In some countries, policy makers have already created and applied policies that have successfully attracted substantial financing to renewable energy, encouraging significant technological advancement alongside massive and rapid deployment. Their experiences provide both inspiration and evidence that the transition to a clean energy system is achievable. Yet, while some countries are moving rapidly in the needed direction, others are still struggling with inertia or have not even begun down this path. A much faster and more global deployment of renewables is required to advance economic development and create domestic jobs, improve energy security, provide energy access to all, reduce local health and environmental impacts and, most important, to reduce greenhouse gas (GHG) emissions dramatically in order to ensure a stable climate.

Getting on track to this sustainable energy future calls for stepped-up policy action starting now. It calls for a focus on advancing renewable energy deployment in combination with major energy-efficiency improvements. Recent international developments, such as the global and regional financial crises, the so-called "Arab Spring," Japan's 2011 Fukushima nuclear accident, and the development of unconventional fossil fuels, have affected circumstances and perceptions surrounding renewable energy. But they have not changed the urgency for change.

Experiences to date point to six key ingredients that policy makers can mix together in their own recipes, adapted to local circumstances, to substantially accelerate deployment of renewable energy—even in the current difficult economic climate—to effectively and efficiently realize the many benefits of a sustainable energy economy.

READy TO GET ON TRACK

This Renewable Energy Action on Deployment publication (READy) publication is intended primarily for use by policy makers. READy is a publication of the International Energy Agency's Implementing Agreement on Renewable Energy Technology Deployment (IEA-RETD), which aims to increase awareness and accelerate deployment of renewable energy technologies. IEA-RETD advocates for expanded international cooperation and public–private partnerships to further the acceleration of renewable deployment.

The current trends in renewables, however impressive, are not on track to keep temperature increases below the internationally agreed 2°C threshold.

READy describes current trends and future outlooks for renewable energy, as well as the barriers that continue to impede broader and more rapid growth. It reviews a number of important energy scenarios to examine potential pathways to a more sustainable energy future.

These are the prelude to the core of the book: the policies that have been proven to work. READy addresses a number of questions to provide guidance for getting on track: Which support policies have been most effective to date and why? Which will be required to drive the transition from a world dominated by fossil fuels to a world in which clean and sustainable renewable energy provides the majority of the world's energy needs? Are any policy options economic crisis-robust? Given the current challenges facing renewable energy—from the global economic slowdown to the rise of shale gas—which specific policy instruments can help to increase significantly the deployment of renewables in the short, medium and long terms?

SIX POLICY ACTIONS FOR ACCELERATED DEPLOYMENT

READy is intended as a resource for policy makers to benefit from lessons learned and a source of inspiration that invites swift action. The inspiration derives from sectoral and cross-sectoral reviews and analyses of policy initiatives and mechanisms, based on experiences to date at the local and national levels. Trends, outlooks, and policies outlined in this report are evidence and fact based, sourced from a variety of sources including numerous other studies commissioned by IEA-RETD.

THE CHANGE NEEDS TO START NOW

Historical patterns of economic development imply that global energy demand will continue to increase over the next few decades. The demand for energy services is increasing sharply in emerging economies like China, India, Brazil, and South Africa. An estimated 1.5 billion people still lack access to modern energy services. Further, the world's population could approach 7.6 billion by 2020, meaning the demands of an additional hundreds of millions of people will need to be met within the next decade alone.

> Without a major transition in the world's energy system, it will be impossible to satisfy all energy needs within the given economic and ecological boundaries.

At the same time, the continued and growing reliance on fossil fuels to meet global energy needs raises significant challenges and costs at the local to global levels. Extraction and burning of fossil fuels enacts high costs on human health and local air and water quality. Concerns related to "peak oil," political instability in key oil-producing regions, and volatile fuel prices pose challenges to countries that are heavily reliant on fossil fuels. And, perhaps most critically, GHGs that are emitted through the combustion of fossil fuels threaten the stability of the global climate.

Without a major transition in the world's energy system—to one based on renewable energy in combination with energy efficiency improvements—it will be impossible to satisfy the needs of current and future populations within the given economic and ecological boundaries. Technologies that harvest biomass, geothermal, hydro, ocean, solar, and wind resources to provide energy services ranging from lighting to mobility to heat have the technical potential to meet global energy demands many times over. Meanwhile, improving energy efficiency is of an equal importance.

SIX POLICY ACTIONS FOR   ACCELERATED DEPLOYMENT

The *International Energy Agency's World Energy Outlook 2011* estimates that for every dollar invested in carbon-intensive energy technologies up to 2020, society will need to pay USD 4.30 to compensate for increased emissions. In other words, if the world postpones action, the 2°C goal will become more expensive to achieve or the costs associated with climate change will increase, or both.

CURRENT TRENDS ARE THE BASIS FOR FURTHER ACCELERATION

Driven greatly by government policies, renewable energy technologies have seen a rapid expansion in deployment and significant reductions in cost in recent years. While this is a good start, the pace must accelerate for the transition to occur in the next few decades, in time to remain below the internationally agreed 2° C threshold in order to avoid catastrophic climate change.

Growth in renewable energy deployment has accelerated in recent years, despite the global economic recession, with notable market growth for many technologies even during 2011. Average annual capacity growth rates for solar photovoltaics (PV) approached 50% from 2005 to 2010 and those for wind energy exceeded 25% during the same period. Rates of growth for renewables in some countries have far surpassed global averages.

The 2011 financial crisis has slowed the growth in some sectors, but has not stopped it. In fact, the world market for wind energy set a new record in 2011, with more than 41 GW of capacity added, while vigorous growth continued in PV markets with the addition of nearly 28 GW. According to data from Bloomberg New Energy Finance (BNEF), global investment in wind power in 2011 was down an estimated 17% from 2010; however, solar PV cost reductions were more than compensated by strong market growth, leading to a 37% increase in PV investments in 2011.

Increasing levels of investment in R&D, manufacturing facilities, and deployment have resulted in steady cost decreases in all renewable energy technologies. Most notable are solar PV technologies, which saw average module costs fall by a factor of almost 50 between 1976 and 2010. Between the summer of 2008 and mid-2011, the price of PV modules declined by an estimated 60% per megawatt, making solar power competitive with the retail price of electricity in many sunny countries. Bio-ethanol and wind energy have also witnessed considerable cost reductions over these time periods, with wind turbine prices falling an estimated 18% between 2009 and 2011.

Renewable energy industry associations foresee continued and rapid market growth in the near future. For instance, the European Photovoltaic Industries Association projected in 2011, under its conservative scenario, that cumulative PV installations would increase by at least 20% annually until 2015, with total capacity more than tripling between year-end 2010 and 2015. Note that actual growth of cumulative capacity in 2011 was approaching 70%. The Global Wind Energy Council predicted that global wind power capacity would more than double during this same period; during the course of 2011, total capacity increased by an estimated 21%. Even so, as baseline capacities increase, relative growth rates will tend to decline, but that the "natural" trend will be offset when renewable technologies become cheaper than fossil technologies.

Government policies have played a crucial role in driving these developments. Between 2004 and early 2011, the number of countries with renewable energy support policies in place doubled to at least 96 countries; more than half of these were emerging or developing economies. Policies include a wide portfolio of measures, ranging from target setting to rebates, to feed-in tariffs (FITs) and quotas.

In addition, local government support schemes are playing an ever-increasing role in advancing renewable energy. Several hundred cities and local governments have adopted their own targets and policies.

Alongside these developments, a growing number of organizations and frameworks have evolved around renewable energy. In addition to traditional players such as environmental organizations and industry groups, stakeholders in the financial community and the business world have established associations that focus on renewable energy, often in combination with energy efficiency. For policy makers, these organizations are an important gateway to a large knowledge database and to new partnerships.

> Recent developments have changed the scene, sometimes, but not always, in a direction that is favorable to renewable energy.

NEW PRIORITIES IN ENERGY INVESTMENT

Global investment in renewable energy increased about sevenfold between 2004 and 2010. In 2011, total global new investment in renewable energy even increased further, reaching USD 257 billion. Growth continued, despite the 2009 and 2011 financial crises, although the picture differs widely according to region and

SIX POLICY ACTIONS FOR   ACCELERATED DEPLOYMENT

technology. In order to get the energy transition on track, however, a larger shift in investment priorities will be required.

During 2010, in the power sector private renewable energy investments in new generating capacity exceeded those in new fossil fuels power plants (including replacement of old plants) for the first time. However, although the increase in public and private investment in renewable energy over the last decade has been impressive, overall global investments in the energy sector still need to change track. Major investments in fossil fuels continue, due in great part to direct subsidies in the form of incentives (such as government payments and tax credits) and indirect subsidies through the failure to price externalities such as carbon dioxide emissions. The more countries invest in existing and new fossil fuel infrastructure today, the more they lock themselves and the world into a high-carbon future and the less they have to invest in a cleaner energy system.

A substantial increase in renewable energy investments has to be achieved in the years ahead, with BNEF projecting up to USD 500 billion annually by 2030.

There are numerous authoritative scenarios that set out possible pathways for achieving large shares of renewable energy, even up to 100% by 2050, thereby reducing GHG emissions substantially. It cannot be denied that the energy transition, like any major change, will require major efforts and investments. But every year of delay results in higher overall costs and further limits the options that are available.

> In 2011, total global new investment in renewable energy reached USD 257 billion.

POLICIES ARE NEEDED TO OVERCOME BARRIERS, CHALLENGES, AND MISPERCEPTIONS

> Some renewable energy technologies are already cost competitive with fossil fuels.

When the urgency for change is so clear and the potential benefits of renewable energy are so large, why doesn't the needed transition occur at the necessary pace and scale? A number of barriers and challenges continue to hold back development and deployment of

renewable energy technologies. Inaccurate perception of the relative costs and benefits of renewable energy is one major barrier; opposition from major players and inertia in the traditional energy system are two other hurdles to overcome.

> The sooner massive deployment occurs, the more (cost) efficient the transition will be.

The scale of renewable energy deployment required over the coming decades is massive. To achieve such advances, a number of barriers and challenges need to be addressed. Most of these barriers are linked to the fact that modern renewable energy technologies are still developing compared to today's more mature fossil and nuclear energy technologies, and to the fact that existing regulations and infrastructure were established to support the existing energy system.

The perception remains that renewable energy is more expensive than fossil fuels. However, some renewable energy technologies are already cost competitive with fossil fuels, and many others would currently be so if existing subsidies for fossil fuels were eliminated and external costs of energy production and use (including climate change and other environmental and health impacts) were included in the price.

Further, most renewable energy technologies face high up-front costs. Although these up-front costs are offset by lower operating costs and reduced price risk for producers and consumers, renewable energy is often perceived by private investors as more risky than fossil fuels. At least a number of cases demonstrate that this perception is inaccurate. But as long as renewable energy projects are perceived as "risky investments," private capital markets will demand relatively high returns

Renewables also face opposition from entrenched and vested interests, which take advantage of their significant market power and political influence to safeguard their positions. Powerful forces and strong lobbies for competing industries work against a transition toward an energy supply based predominantly on renewable energy. Policy makers must find ways to avoid further lock-in to existing infrastructure and technologies, and to overcome the natural inertia against change.

Although studies by IEA-RETD and others show a high overall public support of renewables, specific projects—as with any other type of infrastructure project—can generate local resistance on social, environmental, or

SIX POLICY ACTIONS FOR   ACCELERATED DEPLOYMENT

other grounds, particularly if stakeholder involvement (in planning or ownership) is limited. Such concerns can also pose challenges to deployment and need to be taken into account.

While recent developments on the international stage have added new drivers for renewable energy, others have heightened existing barriers. Oil price volatility (intensified by the Arab Spring), the March 2011 Fukushima nuclear disaster in Japan, major oil spills, and other environmental disasters have further highlighted the costs and risks of the current energy system; at the same time, shale gas discoveries in the United States and elsewhere have reduced natural gas prices and postponed the perceived urgency of developing fossil fuel alternatives.

Most important, the current international economic crisis has changed reality and perception of risk among investors, while also reducing the amount of financing available for renewable energy from traditional sources. Economic and other challenges have also had short-term implications for government policies, and not all of them in ways that are favorable to renewable energy deployment. Following a period marked by numerous green stimulus packages that supported investment in renewable energy, many governments are now reducing support, resulting in further contraction of available financing.

Government policies play a crucial role in lowering or eliminating barriers and challenges to renewable energy deployment. To accelerate the advancement and deployment of renewable energy to the level required over the coming years and decades will require significantly more and stronger policies to support renewable and improve efficiency, in many more countries, combined with aggressive targets and timetables, and a rapid shift away from direct and indirect support for fossil fuels.

> Policies need to adapt to a changed financing landscape by attracting new potential sources of investment

IMPROVING THE RISK-REWARD RATIO TO ATTRACT NEW FINANCING

To attract the needed private investment to renewable energy technologies and related infrastructures, policies need to improve the risk-reward ratio for investors. Substantial financing for deployment projects, manufacturing facilities, and needed

SIX POLICY ACTIONS FOR   ACCELERATED DEPLOYMENT

infrastructure will be critical to getting renewable energy on track in the years to come. At the same time that traditional financing sources are becoming more reluctant to invest due to the global financial crisis, a large amount of private capital is looking for new investment opportunities; renewable energy can provide such an opportunity if predictable, long-term, and effective policies are in place.

> Reducing risk will attract more financing and reduce overall costs.

A large amount of private funding is available for investment, even in the midst of the financial crisis, but renewable energy is competing among a wide choice of investment options. In light of the many investment considerations—amount, level of risk, and expected return—the attractiveness of renewable energy investments must improve and become more visible.

Robust policy support can have the double benefit of reducing risks considerably and, in turn, reducing project and resulting energy costs. Government renewable energy support policies that are long term and transparent, and that create an enabling environment for renewable energy, can improve the risk-to-reward ratio considerably. Recent IEA-RETD work provides many examples of policies, on the national and local levels, that have successfully stimulated considerable growth in renewables, including rebates and other investment incentives, FITs and obligatory renewable shares, clear and simple permitting procedures, low-interest loans, and revolving public funds. A combination of policies is generally required, and the needs will differ from one region or country to another depending on a variety of factors such as available renewable energy resources, electricity generation mix, financing capabilities, public acceptance, and so forth. The overarching factors for success include stability, predictability, and clarity for stakeholders. These factors help to minimize the risk to investors and increase potential benefits, thereby lowering costs and attracting more financing.

As traditional funding sources such as banks tend to shy away from new investment (in renewable and otherwise) policies need to adapt to the changing financing landscape. Policy makers need to adjust existing policies and create innovative new ones to take advantage of possible new financing sources, including pension funds, sovereign wealth funds, insurance funds, private investors, cooperatives, wealthy companies that want to expand into in "green" options, and even recent inventions like "crowdfunding." Now that obligations or bonds are less secure than they once were, such investors are looking for alternative ways to securely invest their money, and renewable energy offers such an option.

> Cities and other local communities have the potential to drive change, not only locally but also on a national or even a global scale.

READy TO LEARN FROM EXEMPLARY POLICIES

There exist numerous examples of policy experiences in a variety of end-use sectors, and across countries and regions, that provide lessons for future policy development. "Carrots" (incentives) and "sticks" (regulations) have been applied in different combinations, in different end-use sectors or economy-wide, revealing which policies and combinations work most effectively and efficiently. Copying and adjusting the good practices to local circumstances will boost the deployment of renewable energy technologies.

The power sector has the longest experience with renewable energy support policies and deployment, but experiences are increasing in the transport and the heating and cooling sectors as well.

In the *power* sector, FITs, if well-designed and implemented, can be most effective and efficient for increasing installed capacity of a range of renewable technologies and for creating broad-based public support. But as seen in Germany, Spain, and elsewhere, design that allows for flexibility alongside predictable adjustments to changing circumstances, such as tariff reductions (or increases) as costs change, are critical to efficiency and success. Quota policies have also played an important role in increasing capacity, particularly of lower cost renewable technologies. At the same time, a combination of policies is needed. Regulations should, for instance, include guaranteed access to the grid, standard rules and minimal connection costs for developers (particularly small scale), and anticipatory transmission planning in preparation for future development.

In the *transport* sector, most countries have used a combination of policy mechanisms, including fiscal incentives such as tax policies, government procurement, trade-related policies, and domestic mandates such as biofuel blending obligations. Of these, mandates focused on biofuels have been most

SIX POLICY ACTIONS FOR ACCELERATED DEPLOYMENT

effective to date. However, they must be paired with efficiency measures and with environmental and social criteria to ensure environmental and social sustainability. Such criteria can be quantified and qualified in terms of CO_2 emissions and other environmental impact, as demonstrated by several business, government, and nongovernmental biofuels initiatives around the world. And finally, early government support of vehicle and infrastructure development is also essential, as seen in Brazil. There has been little experience with policies to promote (renewably powered) electric vehicles, but lessons can be drawn from experiences with biofuels.

Experience is more limited in the *heating and cooling* sector. Fiscal incentives have been used widely in this sector and can be effective, particularly if they focus support on production of heat (or cooling) rather than investment in heating and cooling systems. There has also been success in Spain and elsewhere with obligations combined with quality standards.

For all sectors, a combination of policies is generally needed, including both carrots and sticks. In addition, renewable energy support policies must be part of a broader package of parallel policies that create an enabling environment for renewable energy, for instance, policies that support and define access to the market (electricity grids, pipelines), as well as permitting, spatial planning, and building codes. Priority in policies should be placed on transparency and ease of administration, while minimizing investment risk and attracting capital. Policies must be long term, stable, and predictable to provide the stable conditions required for building market awareness, dealer support, and thus consistent market growth. Some general design elements for renewable energy policies can be identified, including the need for clear long-term targets and transparent plans to achieve them; monitoring, reporting, and subsequent evaluation of progress; and refining targets on the basis of these findings.

Ultimately, the transition to a cleaner, more sustainable energy system will require a change in the energy system as a whole. Therefore, in parallel with the growing share of renewables, modern policy designers are also looking beyond specific end-use sectors. Coevolution of renewable power and electric vehicles is one example of this and designing and implementing smart grids another. Experience with such system transitions or coevolution policies is limited, but evidence is emerging for possible pathways.

Meanwhile, there is growing importance and impact of policies at the local and municipal levels, where policy makers have closer ties to their constituents, and where vested energy interests and industries have minimal influence over policy makers. Inspired by a range of drivers—a secure energy supply, a cleaner local environment and healthier inhabitants, local job creation, and in some cases access to critical energy services—numerous cities and other local communities have adopted targets and enacted policies. The decentralized, modular nature of renewables allows municipalities to deploy renewable at the local level, independent of planning and policies at higher levels of governance. Cities and other local communities have the potential to create change, not only locally but also on the national or even global scale (as seen with Barcelona's Solar Ordinance).

ACTION Star RECOMMENDATIONS

On the basis of the lessons learned, RETD recommends six practical and realistic policy actions for the next five years, which are graphically represented by the six-pointed ACTION Star. Ideally, including a mix of all these actions into the policy portfolio will facilitate a quick start by removing major barriers to deployment and attracting the additional financing required. Policy makers can choose to give priority to one or more particular actions as needed based on historical developments and local circumstances.

IEA-RETD and other studies analyzing experiences with policies for renewable energy reveal three *overarching guidelines* for getting on track:
- Designing a robust financing framework that can withstand economic crises is critical for improving the attractiveness of investing in renewable energy and associated infrastructures within changing landscapes.
- Maximizing benefits and thus support for renewable energy are key. Broad stakeholder participation in planning or ownership of projects can

SIX POLICY ACTIONS FOR   ACCELERATED DEPLOYMENT

minimize opposition (reducing risks and costs), increase available funds for investment, and increase public and political support for renewable energy.

- Solving the inertia and acceptability issues will be a "game changer" as there are powerful forces working against a transition towards an energy supply predominantly based on renewable energy.

Policy makers can yield near-term results by applying six key lessons. Ideally, a policy portfolio contains all six elements (not necessarily in order of priority):

ALLIANCE BUILDING TO LEAD THE PARADIGM CHANGE

The transition to a sustainable energy system based primarily on renewable energy cannot be achieved solely from the top down. A collaborative effort is needed to overcome powerful forces working to maintain the status quo, as well as lock-in to existing infrastructure, technologies, and mindsets, all of which create inertia and slow the pace of change. Collaborative effort among policy makers and stakeholders will also help to ensure that policies are designed and implemented effectively. Thus, new alliances are required among policy makers, investors, environmental organizations, businesses, communities, households, countries, and regions.

COMMUNICATING AND CREATING AWARENESS ON ALL LEVELS

It is important for decision makers across society to recognize the potential, opportunities, and benefits of renewable energy and to have the knowledge, workforce, and skills to realize them. This requires a continuous consultation among stakeholders about their interests and, moreover, adequate communication of benefits and issues of renewable energy.

The broader public (including policy makers) needs to understand the full economic and social costs of the current energy system (including external costs) and the potential of renewables to provide a growing share of energy supply. Misconceptions need to be corrected with adequate information. Policy makers on all levels need to understand what is required to attract investment and to advance renewables so that they can enact strong, long-term and effective policies. Policies and regulations need to be widely known, transparent, and easily accessible to relevant actors. Renewable resource potential needs to be measured and information made accessible. Further, a skilled workforce is required to design, manufacture, install, and maintain renewable energy systems and necessary infrastructure.

TARGET SETTING AT ALL LEVELS OF GOVERNMENT

Ambitious and realistic long-term and interim targets, preferably binding, at different levels of government provide the desired predictability to the energy market. Such targets need to be grounded on clear general goals, and need to be supported by specific renewable support policies that promote not only the deployment of renewable technologies but also the development of needed infrastructures. This will create a stable and favorable investment climate for renewable energy technologies. Countries like Germany, Denmark, and China are proving the effectiveness of ambitious targets in combination with strong support policies, as are many local governments.

INTEGRATING OF RENEWABLES INTO INSTITUTIONAL, ECONOMIC, SOCIAL, AND TECHNICAL DECISION-MAKING PROCESSES

Since the 1980s, many countries have worked to integrate the environment into various areas of policy making and infrastructures; the same should be done with renewable energy, for example, in and across the agricultural sector, urban planning, transportation, and so forth. By incorporating renewables into policies of innovation, finance, spatial planning, and broader economic development, renewable energy will achieve a higher status that will, in turn, help to reduce regulatory inconsistencies and barriers to their deployment. The same applies to the technical integration of renewable energy technologies into grid systems and other infrastructure.

OPTIMIZING OF POLICY INSTRUMENTS

It is important to build on one's own positive experiences, expand those, and learn from the experiences of others. Proven policies can and should be adapted and optimized to local circumstances and needs. In those countries with existing renewable energy support policies, IEA-RETD studies provide evidence that continuing on a consistent policy path—for example, improving an existing policy to address specific concerns and to meet changing

SIX POLICY ACTIONS FOR ACCELERATED DEPLOYMENT

circumstances—is generally more effective than switching policies midstream. This is because investors gain confidence from stable and predictable policies.

  # NEUTRALIZING OF DISADVANTAGES AND MISCONCEPTIONS

Energy markets are built on government rules and (financial) interventions, many of which are not consistent with the goal of a low-carbon energy system, and renewables are trying to compete on an uneven playing field. Moreover, misconceptions exist about costs and reduction perspectives of renewable energy, among other issues.

Developing an inventory of existing support mechanisms for fossil fuels, such as direct subsidies or tax reductions, can lead to identification of distortions in the playing field. Integrating external costs (and benefits), such as climate change or health impacts, into energy prices will also help to level the playing field for renewables, while also revealing the real costs and benefits associated with increasing deployment of renewable energy. Pricing carbon by taxing CO_2 or establishing a CO_2 emission allowances market are examples of such policies.

Getting on track toward a sustainable renewable future at the pace required calls for ACTION now!

ROADMAP TO READy

This book aims to inspire and guide decision makers, and particularly policy makers, to lay a sound foundation for effective and efficient, substantial, and rapid deployment of renewable energy. The inspiration, practical information, and success stories for policy makers in this book are derived from existing IEA-RETD studies and many other sources.

READy consists of the following three parts.

Part I (Trends and Outlooks) aims to establish a baseline for renewable energy and answer these questions: How far have renewable technologies come? What is their current status? It also provides possible scenarios for the role renewables could (and should) play by 2050 in meeting global energy supply and addressing climate change. Part I describes recent trends and the current status of renewable energy technologies, including recent market, industry and investment developments, cost reductions, and increasing shares

SIX POLICY ACTIONS FOR   ACCELERATED DEPLOYMENT

of renewables in energy demand. A number of scenarios are reviewed, including the RETD Outlook, as well as a progression of the IEA *World Energy Outlook* scenarios, an overview of scenarios reviewed in the recent IPCC *Special Report on Renewable Energy and Climate Change Mitigation*, and two scenarios that present possible pathways to (near) 100% renewable energy by 2050. It goes on to discuss the primary barriers to renewable energy and issues to consider related to renewables and sustainability.

Part II (Policy Experiences and Lessons Learned) aims to answer the question: What policies have been used to advance renewable energy and what lessons can be learned from experiences to date? It examines challenges and potential policies for attracting the needed financing, and sets out policy options for advancing renewable energy in each end-use sector and for the energy system as a whole. Throughout Part II, there are inspirational examples and best cases that discuss lessons learned that are transferable to other locations and situations.

Part III (The Road Ahead) concludes the book with key policy recommendations for the road ahead, brought together in the six-pointed ACTION Star. The ACTION Star is a simple guide that policy makers can use to design a portfolio of policies in order to achieve rapid deployment of renewable energy while also building a robust foundation for long-term developments.

No single policy blueprint will guarantee success, nor will readers be able, apart from very special occasions, to copy-and-paste successful policies for use in their own countries and circumstances. In most cases policies and portfolios must be adapted or combined with other policies to improve the odds for success. However, learning from the best practices elsewhere will help decision makers in government, business and the broader community to accelerate the transition towards a clean, sustainable energy system.

SIX   POLICY   ACTIONS   FOR   ACCELERATED   DEPLOYMENT

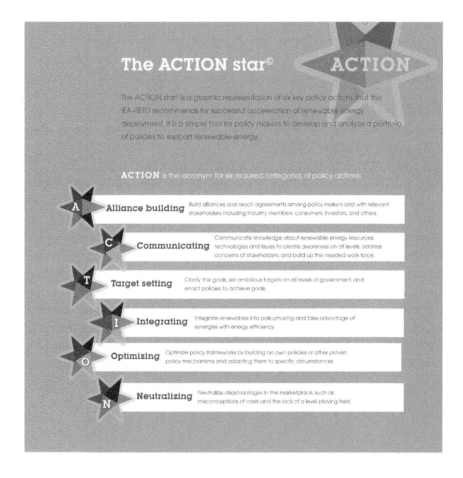

The ACTION star©

The ACTION star is a graphic representation of six key policy actions that the IEA-RETD recommends for successful acceleration of renewable energy deployment. It is a simple tool for policy makers to develop and analyze a portfolio of policies to support renewable energy.

ACTION is the acronym for six required categories of policy actions:

Alliance building — Build alliances and reach agreements among policy makers and with relevant stakeholders including industry members, consumers, investors, and others.

Communicating — Communicate knowledge about renewable energy resources, technologies and issues to create awareness on all levels, address concerns of stakeholders, and build up the needed work force.

Target setting — Clarify the goals, set ambitious targets on all levels of government, and enact policies to achieve goals.

Integrating — Integrate renewables into policymaking and take advantage of synergies with energy efficiency.

Optimizing — Optimize policy frameworks by building on own policies or other proven policy mechanisms and adapting them to specific circumstances.

Neutralizing — Neutralize disadvantages in the marketplace, such as misconceptions of costs and the lack of a level playing field.

SIX POLICY ACTIONS FOR ACCELERATED DEPLOYMENT

LIST OF CASE STUDIES

FIGURES

TABLES

PART ONE

Trends and Outlooks

CHAPTER ONE

Global and Regional Trends in Renewable Energy

Contents

Renewable energy is clearly entering the mainstream. In 2010, an estimated 16.7% of global final energy was supplied by renewable energy—including hydropower, wind, solar, geothermal, and traditional and modern biomass energy. Modern renewable energy provided about 8.2% of this total, with traditional biomass accounting for an estimated 8.5% according to the _REN21 Renewables 2012 Global Status Report_. Further, renewables accounted for almost half of new electric capacity added worldwide during 2011, and generate about 20.3% of global electricity based on generating capacity in operation at year's end. Market growth has been rapid in recent years, with an accelerated growth rate for some technologies over the past half-decade. The global economic crises of 2009 and 2011 slowed growth in some technologies and regions, but overall growth has continued.

Renewable energy markets have continued their rapid growth in recent years, both in total scale and in geographic distribution, in spite of the global economic recession. Global investment in renewable power and fuels, at all stages of development, has also risen rapidly and was up more than sixfold between 2004 and 2011. In response to technology advances

Renewable Energy Action on Deployment
http://dx.doi.org/10.1016/B978-0-12-405519-3.00001-3
3

and greater economies of scale, renewable energy costs have fallen dramatically and further cost reductions are expected in all technologies. All of these developments are due greatly to a significant increase in government support policies for renewable energy and a growing number and expanding breadth of international players involved with renewable energy.

This chapter reviews the major trends in renewable energy development over the past two decades, with emphasis on the past five years. It begins with an overview of market trends by the end-use sector and technology, followed by developments in renewable energy investment in recent years, and past cost trends and future expected reductions. Trends related to government policies focused specifically on renewable energy are covered next, followed by a brief discussion of relevant international players in the renewable energy arena, ranging from international organizations to international finance organizations to nongovernmental organizations (NGOs) and the private sector. The final section provides an overview of various projections for future market development and renewable energy deployment.

Except where otherwise noted, data and trends in this chapter are drawn primarily from the *Renewables 2012 Global Status Report* of REN21 (Renewable Energy Policy Network for the 21st Century); the investment data are from Bloomberg New Energy Finance (BNEF) and drawn from *Global Trends in Renewable Energy Investment 2012* (UN Environment Programme, Frankfurt School, and BNEF).

1.1. RENEWABLE ENERGY MARKETS

Renewable energy markets have grown rapidly over the past several years, particularly in the electricity sector, where solar photovoltaics (PV) has witnessed rapid expansion and has remained the world's fastest growing energy technology for some time. Figure 1.1 shows growth rates of several renewable technologies for the year 2011, and annual averages for the five year period from the end of 2006 through 2011. Many technologies experienced remarkable growth, with solar PV and solar hot water/heating seeing even higher rates of growth during 2011 than the average over the 2006–2011 period.

After PV, concentrating solar thermal power (CSP) has witnessed the most rapid relative growth in the electricity sector over the past five years, growing from a small base, and wind power capacity has also grown rapidly. In terms of total new capacity added during this period, wind power takes the lead with about 164 GW, followed by hydropower. As seen in Figure 1.1, biofuel production and consumption took off in the middle of the last decade, but growth rates have since slowed, with ethanol production remaining stable

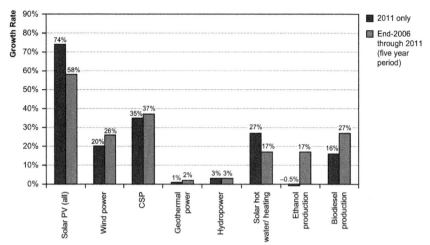

FIGURE 1.1 Average Annual Growth Rates of Renewable Energy Capacity and Biofuels Production, 2006–2011 *(From REN21, Renewables 2012 Global Status Report).* For color version of this figure, the reader is referred to the online version of this book.

or declining slightly in 2011. The renewables heat sector is considered by many to be a sleeping giant, with enormous potential that is still waiting to be tapped, although markets have begun to pick up in recent years. Table 1.1 shows the changes in cumulative capacity or annual production that various renewables witnessed between 2006 and 2011.

> In 2011, renewable energy accounted for almost half of the electric capacity added worldwide.

1.2. ELECTRICITY MARKET TRENDS: RAPID GROWTH

In 2011, renewable energy accounted for almost half of the ~208 GW of electric capacity added worldwide. Renewables account for about a quarter of the world's electric capacity (estimated to be 5360 GW at the end of 2011) and provide an estimated 20.3% of global electricity supply, based on total global generating capacity in place at the end of 2011.

Hydropower

Hydropower is used in more than 150 countries and generates most of this supply, accounting for an estimated 15.3% of global electricity production in 2011. Asia leads the world for total hydropower capacity, followed by Europe, North and South America, and finally Africa at a distant fifth. Asia

TABLE 1.1 Global renewable energy installed capacity or annual production, 2006 and 2011

Technology	2006 (GW)	2011 (GW)
Hydropower[a]	~830	970
Wind power	74	238
Biomass power	45	72
Solar power		
• PV	7	70
• CSP	0.4	1.8
Geothermal power	10	11.2
Ocean power	0.3	0.5
	GW_{th}	**GW_{th}**
Biomass heat	NA	290
Solar heat[b]	107	232
Geothermal heat	33	58
	billion liters per year	**billion liters per year**
Biofuels		
• Ethanol	39	86
• Biodiesel	6	21

[a]Hydropower data are estimated and do not include pure pumped storage.
[b]Solar heat data are estimated and do not include unglazed systems (for swimming pools).
Source: Data rounded and sourced from REN21, *Renewables Global Status Report*, various years.

and South America, led by China and Brazil, are the most active markets for new development, with several large dams under construction as well as many small-scale projects. New plants are also being constructed in North America and Europe, although the greatest focus is on modernization of existing plants and development of pumped storage to support high-capacity shares of variable renewable technologies.

Despite the large increases in hydropower capacity in recent years, as seen in Table 1.1, hydropower's share of total global electricity production has declined (from 19% in the mid-1990s) as consumption has expanded at a faster pace. Further, hydropower's share of total renewable capacity is falling as markets for other renewable technologies surge. In 2004, hydropower accounted for about 89% of total installed renewable electric capacity; by the end of 2011, its share had fallen to an estimated 71.3%.

Wind Power

After hydropower, wind power has experienced the largest capacity gains, rising rapidly from 74 GW in operation at the end of 2006 to nearly 238

GW at the end of 2011. An estimated 40 GW was added during 2011, increasing total global capacity by about 20% relative to year end 2010.

By the end of 2011, wind power capacity installed was enough to meet an estimated 2–3% of global electricity consumption; while still a small share, it was up significantly in a few short years. And wind power's share of total demand was far higher in several individual countries—for example, nearly 26% in Denmark, 15.9% in Spain, 15.6% in Portugal, and 12% in Ireland—while wind power capacity installed in the European Union (EU) by the end of 2011 was enough to meet 6.3% of the region's electricity use in a normal wind year (up from 5.3% in 2010). Shares are even higher in many states and provinces. For example, four German states met more than 46% of their electricity demand with wind in 2011; South Australia produces 20% of its electricity with wind; and five U.S. states met more than 10% of their electricity demand with wind in 2011.

Policy uncertainty and the continuing economic crisis have combined to slow the growth of wind power in developed countries, although the United States saw a jump in capacity additions in 2011 driven by the approaching expiration of some important federal incentives. For the second year in a row, the majority of new capacity was added in developing countries and emerging markets rather than in wind's traditional markets. Leadership has shifted from Europe (initially Denmark, and more recently Germany and Spain) to the United States, and most recently to China, which accounted for half the global market in 2010 and an estimated 44% in 2011. In recent years, wind power has become one of the broadest-based renewable energy technologies, with commercial installations in more than 70 countries by 2006, and in 83 countries by 2010; at least 68 countries around the world had more than 10 MW of reported wind power capacity in 2011.

Offshore wind power shows great promise, but to date markets have developed relatively slowly due to high costs and siting and permitting challenges. Most capacity is installed off the coastlines of Europe; the first major offshore wind project outside of Europe began operating in China during 2010. There is a trend toward increasing the size of individual wind turbines as well as the capacity of individual projects; at the same time, interest in community-based wind power projects is on the rise beyond the traditional cooperatives in Denmark and Germany, and interest in small-scale turbines is also increasing.

Biomass Power

Trends in biomass power markets are more difficult to track because of the diversity of feedstock and technologies, and because there are no industry associations that undertake a systematic and comprehensive compilation of global data (as there are for wind and solar PV, for example). However, it is clear that the use of biomass for electricity generation is increasing, particularly in Europe (led by Germany, Sweden, and the UK), the United States, Brazil, China, India, and Japan. Although on a much smaller scale, biomass power has also grown significantly in several other Latin American and Asian countries, including Costa Rica, Malaysia, Mexico, Thailand, and Uruguay, and there is increasing interest in modern biomass power in Africa as well.

Global biomass power capacity reached an estimated 72 GW in 2011. Approximately 88% of biomass power is generated with solid biomass fuels, with the remainder provided by biogas and liquid biofuels.

In many countries, much if not most biomass power capacity is in the form of cogeneration (or combined heat and power; CHP), particularly in some European countries and in countries with large sugarcane industries, where waste bagasse is used to produce heat and power. Solid biomass is also being used increasingly for cofiring with fossil fuels (particularly coal) in existing power plants. Use of biogas (from agricultural sources or landfill gas) is also expanding rapidly; for example, most biomass power in Germany came from biogas as of 2010, with capacity increasing more than 20% during that year alone.

Solar PV

Almost 30 GW of solar PV capacity came into operation during 2011, bringing total global capacity to nearly 70 GW by year's end. This represents about 10 times global capacity in operation only five years earlier. Cumulative global PV capacity increased at an average annual rate of more than 58% between 2006 and 2011, and grew 74% in 2011.

The EU dominated the global PV market in 2011, as it has for the past several years, accounting for about 57% of new operating capacity. Italy took the top spot from Germany in 2011, followed by China (moving up from sixth in 2010), the United States, France, and Japan. Germany, which led the market for five years, added more capacity during 2010 than the entire world installed in the previous year; by the end of 2010, Germany alone had more total capacity than the world did two years earlier. As of the end of 2011, Germany had an estimated 24.8 GW of PV capacity in operation, followed by Italy with 12.8 GW.

Since 2005, when nearly all global PV systems were installed remotely and off the grid, the use of PV has shifted dramatically and almost all capacity was grid connected by the end of 2011 (although off-grid capacity continues to rise). While most of the world's PV systems are still small scale and distributed, there is a trend toward utility-scale PV plants. By early 2012, at least 12 countries (in Europe, North America, and Asia) had plants exceeding 20 MW.

Although the concentrating PV market is still small, interest is increasing in this technology as well. During 2011, the world's first multi-megawatt projects were installed. The largest markets by year's end were Spain and the United States, but projects were operating in at least 20 other countries around the world.

CSP

The CSP market has also been focused on Europe (primarily Spain) and the United States. After several years of inactivity, capacity increased from less than 400 MW in 2005 to an estimated 1,760 MW at the end of 2011, with more than 450 MW added during 2011. Although Spain and the United States remain home to almost all of the world's existing capacity, projects are now under construction or planned in several countries in North Africa, the Middle East, and Asia. Egypt and Morocco brought their first projects on line in 2010, and Algeria, Thailand, and India started operating their first CSP plants during 2011.

Due to rapid reductions in PV costs, in 2010 and 2011 several planned projects in the United States shifted from CSP technologies to utility-scale PV. In addition, the Arab Spring slowed development in the Middle East–North Africa region during 2011. However, CSP growth is expected to accelerate internationally in the coming years.

Geothermal Power

Geothermal power markets have been relatively slow in recent years, increasing at an average annual rate of just over 2% between 2006 and 2011. More than 135 MW came on line during 2011, with most of this in Iceland, and the rest in Nicaragua and the United States. Geothermal electricity generation was an estimated 69 TWh in 2011, and global capacity reached approximately 11.2 GW by year's end.

Geothermal power plants are operating in at least 24 countries, with the vast majority of global capacity located in the United States, the Philippines, Indonesia, Mexico, Italy, Iceland, New Zealand, and Japan. Iceland produced about 26% of its electricity with geothermal energy in 2010, and the Philippines generated approximately 18% of its total. Kenya, with about 200 MW of capacity in operation, leads Africa for geothermal electric capacity.

Deployment is expected to accelerate as markets broaden across Africa, Latin America, and elsewhere, and as advanced technologies enable development of geothermal power in new locations.

Ocean Energy

Ocean energy technologies are still relatively immature, but they continue to advance. After years with development of only small pilot ocean energy projects, global capacity almost doubled in 2011 with the addition of about 254 MW. Almost all of this capacity began operation off the shores of South Korea, and it brought total global ocean energy capacity (most of which is tidal power) to 527 MW. Elsewhere around the world, numerous projects of various scales, from small-pilot plants to utility-scale projects, are in development or under contract. At least 25 countries are actively developing projects, particularly wave and tidal plants.

1.3. TRANSPORTATION MARKET TRENDS: A MIXED PICTURE

Renewable energy is used in the transportation sector in the form of biofuels, biogas, renewable electricity for electric vehicles, and renewably produced hydrogen. Data are available at the national level for some of these options; for example, biogas accounted for 11% (on an energy basis) of the total 5.7% biofuels share of Sweden's transport fuels in 2010. However, global market and production data are available only for liquid biofuels. Production of ethanol and biodiesel increased rapidly from 2000 to 2010, particularly in the middle of the decade. Growth has since slowed, with ethanol production holding steady or dropping slightly during 2011, and biodiesel increasing by about 16% relative to 2010.

Most renewable fuel is used for road transport, but there is increasing interest in aviation biofuels, and several airlines around the world began operating commercial flights with various biofuel blends during 2011.

Fuel Ethanol

The United States and Brazil lead the world for ethanol production, accounting for 87% of the 86 billion liters produced in 2011. After more than 25 years as the leading producer and user of ethanol, Brazil was overtaken by the United States in 2006. In 2010, after several years as a net importer, the United States became a net exporter and began to take international market share from Brazil, particularly in the traditional markets in Europe. This trend continued in 2011, as ethanol production fell nearly 18% in Brazil due greatly to declining investment in sugarcane production, poor harvests due to unfavorable weather conditions, and high world prices for sugar. The growth of the U.S. market is a relatively recent trend—U.S. ethanol production capacity rose from only 4 billion liters per year in 1996 to 14 billion liters per year capacity in 2004, and in 2011 the country produced a total of more than 54 billion liters.

Biodiesel

Global biodiesel production increased from less than 1 billion liters per year in 2000 to more than 21 billion liters per year in 2011; most of this growth occurred between 2005 and 2008. Biodiesel production is far less concentrated than ethanol. After several years as the world's leading producer, Germany was overtaken in 2011 by the United States, which increased production by almost 160%, to almost 3.2 billion liters. The United States was followed closely by Germany, with Argentina coming in third (due to a 34% increase over 2010), and Brazil ranking fourth globally. The EU remained the largest regional producer in 2011, but its share of world production is declining.

1.4. HEATING/COOLING MARKET TRENDS: SLOW AWAKENING OF THE "SLEEPING GIANT"

Biomass, solar, and geothermal resources can provide heat for a variety of purposes, ranging from space and water heating to cooking to process heat for industrial uses. Renewable resources can also provide air conditioning and cooling, but technologies and markets remain relatively immature and little data are available. Compared to the potential, the growth trends in renewable heat remain modest.

Biomass Heat

Globally, biomass (whether modern or traditional) provides the majority of heating produced with renewable energy. An estimated 10 GW_{th} of modern biomass heating capacity was added during 2011, bringing worldwide use of modern biomass for heat production to 290 GW_{th}. Heat is produced through the burning of solid, gaseous, and liquid biomass, and can be used for water and space heating, industrial and agricultural processes, and cooking.

The use of biomass for district heat and combined heat and power has been expanding in a number of countries, particularly in Europe. In addition to solid biomass, biomethane (purified biogas), which can be injected into the natural gas grid, is contributing to heat or power production. It has been used for decades in Europe, but production and use have recently picked up speed due, in great part, to rapid expansion in Germany.

Biomass pellets are also becoming an increasingly common fuel for heat production, especially in the EU. They are burned in small units to heat residential or commercial buildings as well as in CHP plants.

Among developing countries, small- to large-scale CHP plants are becoming more common—they rely greatly on agricultural residues such as coconut or rice husks, as well as bagasse residues from sugar production. The use of small-scale biogas plants, for cooking and other purposes, is also increasing, particularly in China and India.

Solar Heating and Cooling

Solar heating technologies are also becoming more widespread, although China continues its long-term position as global leader in the manufacture and use of solar water heating equipment. Excluding unglazed systems (for swimming pool heating), an estimated 232 GW_{th} of solar heat capacity was installed worldwide by the end of 2011, more than double the 2006 total of just under 110 GW_{th}.

While Europe (especially Germany) continues to represent a significant portion of the global market, many EU countries have lost market share in the past few years due to policy changes or falling natural gas prices, which have reduced the cost-competitiveness of solar for heating. In the meantime, rapid market expansion is underway in Brazil and India, and interest is picking up elsewhere in Latin America and Asia, as well as in several African countries, although markets are still very small.

While most solar thermal technologies are used to heat water, there are growing markets for other purposes. Combination systems, which provide water and space heating, are gaining ground in some EU countries, as are systems that can also provide air conditioning and cooling.

Interest in solar heating and cooling is spreading to the Middle East and Asia as well, and a limited but growing number of solar process heat systems were in operation or under construction from Europe to China to South Africa by early 2011. Global data are not available for installed capacity of solar cooling technologies.

Geothermal Heating and Cooling

The use of geothermal to produce heat is also expanding and spreading around the globe. Geothermal direct heat represents about two-thirds of the estimated 205 TWh (738 petajoules) of geothermal energy generated during 2011. Use nearly doubled from 2000 to 2005, and heat output from geothermal sources increased at an average annual rate of 10% between 2005 and 2010. At least 78 countries used direct geothermal energy in 2011, up from 58 in 2000.

Ground-source heat pumps, which represent about 72% of global geothermal direct heat capacity, have seen rapid growth and account for most of the increase in geothermal heat over the past decade. Most of the installations have occurred in North America, Europe, and China. Heat pumps can also be used for cooling purposes.

Other uses of direct geothermal energy include bathing and swimming applications (almost 25% of geothermal direct energy use); direct space heating, mainly with district heat systems (14%); and heat for greenhouses, aquaculture pond heating, agricultural drying, industrial purposes, cooling, and snow melting. Iceland met about 90% of its heating demand with geothermal resources in 2011 and ranked first in the world for average annual energy use per person from geothermal, followed by Sweden, Norway, New Zealand, and Denmark.

1.5. INVESTMENTS INCREASING AT ALL DEVELOPMENT STAGES

Investment in renewables occurs in each of the stages from research and development (R&D) through to commercialization and beyond. These five stages include: R&D, technology development and commercialization, equipment manufacture and sales, project construction, and refinancing and sale of companies, mostly through mergers and acquisitions. Over the past several years, investments have increased in each of these stages.

Total New Investment

In 2011, total global new investment in renewable energy reached USD 257 billion, an increase of 17% relative to 2010, according to financial analyst Bloomberg New Energy Finance (BNEF). This figure includes: all biomass and waste-to-energy, geothermal, and wind generation projects of more than 1 MW; all hydropower projects of between 1 and 50 MW; all wave and tidal energy projects; all biofuel projects with a capacity of one million liters or more per year; and all solar projects, with those less than 1 MW estimated separately. This increase in investment occurred during a time of economic and policy uncertainty in a number of countries, particularly in the developed world, and rapidly declining prices for some technologies. BNEF estimates that the increase in investment would have been significantly larger in 2011 if not for declining costs of wind and solar PV technologies.

Not included in BNEF's estimates are investments in solar hot water collectors, estimated to be at least USD 10–15 billion in 2011, and an additional USD 25.5 billion or more invested in hydropower projects larger than 50 MW. Including these investments, total global new investments in renewable energy in 2011 exceeded USD 292.5 billion.

Investment by Type

Of the major investment types, significant increases were seen in asset finance of utility-scale projects in both 2010 and 2011, and in small-scale distributed capacity (particularly rooftop solar power), which increased by 18 and 25%, respectively, in 2011. Both reached record highs in 2011. After rising in 2010 due greatly to green stimulus funds that were established in response to the global economic crisis, government-funded R&D declined in 2011, as did private investment in R&D. Declines were also seen in private equity expansion capital investment, and in equity raising by renewable energy companies on public markets.

Electric Generating Capacity

Investment in electric generating capacity came to an estimated USD 187 billion in 2010 for large- and small-scale renewables (excluding hydropower projects larger than 50 MW), and to USD 233 billion including all hydropower. In contrast, BNEF estimates that global gross investment in fossil fuel power capacity was USD 219 billion during 2010 (with net investment totaling USD 157 billion). Including all technologies, renewables investment in new generating capacity exceeded fossil fuels in terms of both gross and net investment in 2010. In 2011, net investment (which also covers replacement plants) in renewable power capacity, including all hydropower, was an estimated USD 262.5 billion, approximately USD 40 billion higher than the same measure for fossil fuels.

Investment Dollars

A longer term view of renewable energy investment and financing trends shows that the USD 257 billion in 2011 compares with USD 39 billion in 2004, the earliest year for which such numbers are available. As seen in Figure 1.2, growth has been fairly steady over the intervening years—with

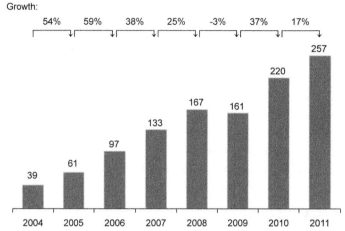

Growth:

54%   59%   38%   25%   -3%   37%   17%

FIGURE 1.2 Global New Investment in Renewable Energy, 2004–2011. *(From BNEF from UNEP/Frankfurt School/BNEF, Global Trends in Renewable Energy Investment 2012; Frankfurt: UNEP Collaborating Centre, Frankfurt School of Finance and Management, 2012).* For color version of this figure, the reader is referred to the online version of this book.

the exception of 2009 when, in the wake of the global financial crisis, investment declined slightly relative to 2008 levels—averaging 31% annually during the 2004 to 2011 period.

> Annual growth of investments amounted to an average of 31% during the 2004–2011 period.

Leading Countries

In 2010, the top countries for total investment were China—which with close to USD 50 billion accounted for nearly one-quarter of the global total and led the world for the second year in a row–followed by Germany (USD 41 billion), the United States (USD 30 billion), Italy (USD 14 billion), and Brazil (USD 7 billion). China maintained its lead in 2011, with investment increasing by 17% to an estimated USD 52 billion. But the United States was not far behind, with an increase of 57% during 2011 to USD 51 billion, driven greatly by the pending expiration of three key federal support policies. Developers rushed to finance projects to take advantage of U.S. incentives while they were still available. Including investment in energy efficiency, smart grids and related technologies, the United States was the leading investor in 2011. Investment in Germany declined in 2011 compared with 2010 but continued to rank third globally, followed by Italy. India saw an increase of 62% and came in fifth, with investment of USD 12 billion during the year; the increase reflected a sharp rise in financing of solar projects under the National Solar Mission, as well as a rise in additions of wind power capacity, growth in venture capital, and an increase in private equity investment in renewable energy companies. Brazil also saw an increase in investment to primarily wind power.

After several years of consistent increases in the share of total global investment, developing countries accounted for approximately 35% during 2011, compared with 65% for developed countries.

Leading Technologies

Another important trend is the rise of solar power, particularly solar PV. As recently as 2006, investments in wind power were twice those in solar power; in 2010, wind investment (USD 90 billion) was only 14% higher than solar (USD 79 billion) and, in 2011, solar power investments (USD 147 billion) jumped 52% to almost double the total investment in wind power (USD 84 billion, down 12% relative to 2010).

BNEF has attributed much of the increase in solar PV investment to falling solar panel prices and solar feed-in tariffs (FITs), which have driven rooftop PV markets in several European countries—an estimated 86% of small-scale solar PV investment during 2010 occurred in countries with FITs.

Biomass power was the third largest sector for total investments during 2011, following solar and wind power. Biofuels ranked fourth, but saw a decline of 20% during 2011 compared with 2010 investment levels.

Small-scale projects, in general (dominated by solar PV because renewable heat systems are not included in BNEF numbers), have experienced dramatic growth in recent years, increasing more than sixfold from USD 9 billion in 2004 to USD 60 billion in 2010, and accounting for more than half of the rise in overall global investment in 2010. Despite rapidly declining prices for solar PV, global investment in small-scale projects increased to USD 76 billion in 2011.

Investors

Renewable energy investments have come from an ever-more diverse range of private and public sources. It is only within the past decade that large commercial banks have begun to sit up and pay attention to renewable energy; in addition, traditional utility companies across the United States and Europe have started to make serious investments in wind power and solar PV. Other large investors include leading investment banks and venture capital investors.

State-owned multilateral and bilateral development banks have become increasingly important in the past few years due to the global economic crisis, providing asset finance for renewable energy. According to BNEF, more public money was directed to renewables (USD 15.2 billion, mostly as loans) through development banks than through government stimulus packages during 2010. The year 2011 saw an even larger commitment, of USD 17 billion, amounting to about four times the 2007 total.

The three leading development banks, in terms of financing in 2010, were the European Investment Bank (USD 5.4 billion), BNDES of Brazil (USD 3.1 billion), and KfW of Germany (USD 1.5 billion), all of which have seen investment increases since 2007. The EIB (USD 4.8 billion) and BNDES (USD 4.6 billion) remained the top two providers of finance among the development banks in 2011. Funds from the World Bank Group directed specifically to renewable energy have increased from an average USD 110 million annually in 2002–2004 to USD 748 million in 2010. At the same time, however, the WBG, which along with KfW and the Global Environment Facility was one of the top sources of funds in 2004, provided only 14% as much as the EIB did during 2010.

Local sources of financing have also been growing, particularly in developing countries, with industry associations, NGOs, international partnerships and networks, as well as private foundations playing increasingly important roles.

R&D Funding

In 2010, for the first time ever, governments spent more on R&D than the private sector did. This was thanks to green stimulus money, particularly in Australia, Japan, and South Korea. Governments invested USD 9 billion in R&D in 2010 compared with the private sector's USD 3 billion (down from USD 4 billion in 2009). Most of this went to solar (USD 3.6 billion), followed by biofuels (USD 2.3 billion). In 2011, however, government R&D fell by 13% to USD 4.6 billion and corporate R&D dropped by 19% to USD 3.7 billion.

Threats to Future Investment

Although expansion in renewable energy capacity and investment has continued to grow steadily (with the exception of a slight dip in 2009) over the past several years, BNEF warns that there is no guarantee of a smooth continuation in the next few years.

Economic problems that have affected the sector since 2008 continued to pose a threat. In particular, the Euro area sovereign debt crisis began affecting the supply of debt for renewable energy projects in Europe by the end of 2011, as banks responded to increases in their cost of funding and upgraded their assessment of risks associated with lending in some EU countries. A number of European governments have struggled to adjust payments (particularly for solar PV) under existing feed-in laws, primarily in response to falling prices, which have dropped faster than costs; the resulting installation booms have led to further reductions in support. In addition, Congressional support in the United States has declined due to a drop in natural gas prices. More broadly, governments have become more reluctant to enact policies that would increase energy prices at a time when consumers are under increasing financial pressure.

BNEF notes that such reductions in policy support might seriously affect renewable energy investment in developed countries during the years from 2012 to 2014, at a time when fully competitive renewable power could be a few short years away. The result could be further challenges for businesses in the industries (particularly solar and wind power) and provide higher expectations hope for limiting carbon emissions in the electricity sector.

1.6. COST TRENDS AND FUTURE PROSPECTS

The so-called "levelized cost of energy" (LCOE) for renewable technologies varies quite widely, depending on technology characteristics, regional variations in cost and performance, local costs of operation and maintenance, and applicable discount rates. Although LCOE for some renewable technologies is still higher than current *energy market prices* for fossil energy sources, renewable energy is already economically competitive in many locations and circumstances and costs continue to fall.

It is important to note that renewable energy generally presents far lower "external costs" (e.g., environmental, health, security, economic costs that are not accounted for in market prices) than do fossil fuels. If the external costs and benefits associated with energy production and use were internalized—or incorporated into *energy market prices*—the competitiveness of renewables would be further improved. In the United States alone, external costs (e.g., health and agriculture damage, economic disruptions from oil price shocks; excluding climate change damage) were estimated to be at least some USD120 billion in 2005. In contrast, renewables provide broad economic, environmental, and social benefits that are generally not monetized. Furthermore, the attractiveness of a renewable energy option also depends on the contribution that a technology can make to meeting specific needs (such as electricity to meet peak demand), or added costs (for integration, etc.) that it imposes on the energy system.

In addition, historically and currently, the lion's share of government subsidies around the world has been directed to fossil and nuclear energy, rather than to renewables. In its annual *World Energy Outlook 2011,* the International Energy Agency (IEA) estimated that in 2010 fossil fuels received about USD 409 billion in subsidies globally, compared with USD66 billion provided to renewable energy sources.

While the general trend in costs of energy from renewable sources has been downward, in recent years there have been relatively brief periods of rising prices for some technologies due to rapid increases in demand that outstrips available supply—either for renewable energy technologies or for input materials. For example, installed costs of wind energy projects rose between 2004 and 2009 due to a variety of factors, including an increase in costs of material inputs, higher labor costs, and currency fluctuations. Solar PV costs increased temporarily in the middle of the last decade due to global shortages of polysilicon, combined with strong demand growth (in excess of supply).

Nevertheless, significant cost reductions have been experienced by all renewable technologies as a result of ongoing public and private R&D economies of scale in production; increased experience with installation, operation, and maintenance; and increased market competition.

According to the IEA report *Deploying Renewables 2011,* considerable cost reductions continue to occur. The report notes that "a portfolio of renewable energy technologies is becoming cost-competitive in an increasingly broad range of circumstances, in some cases providing investment opportunities without the need for specific economic support."

Price Reductions to Date

Figure 1.3 shows experience curves for silicon PV modules (global, per unit of capacity), onshore wind (in Denmark, per unit of capacity), and sugarcane-based ethanol (in Brazil, per unit of production). According to the Intergovernmental Panel on Climate Change (IPCC) *Special Report on Renewable Energy Sources and Climate Change Mitigation* (SRREN), reductions in costs or prices per unit of capacity understate reductions in LCOE of a technology when improvements in performance occur. However, these curves clearly show the trend of declining prices that these technologies have experienced over the past few decades.

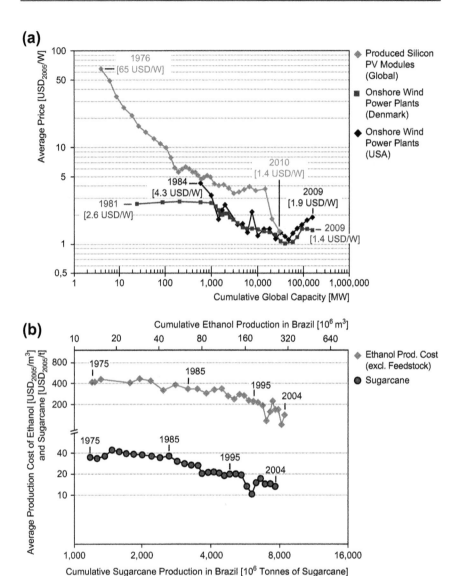

FIGURE 1.3 Experience curves (logarithmic scale) for (a) the price of silicon PV modules and onshore wind power plants per unit of capacity and (b) the cost of sugarcane-based ethanol production *(From: IPCC 2011: Special Report on Renewable Energy Sources and Climate Change Mitigation, Summary for Policymakers, Figure SPM.6. Cambridge University Press).* For color version of this figure, the reader is referred to the online version of this book.

The IPCC estimates that, in the past ten years, the costs of solar water heaters in Europe have declined 20% for each 50% increase in installed capacity. Further, it estimates that PV prices have fallen by more than a factor of 10 in the past three decades—with module prices declining from USD 22/W (measured in 2005 USD) in 1980 to 2005 USD 1.50/W in 2010, and balance of system costs also declining steadily.

According to BNEF, the price of solar PV modules actually declined 60% per megawatt between the summer of 2008 and mid-2011, putting "solar power on a competitive footing with the retail price of electricity in a number of sunny countries"—a position that is sometimes called "grid parity". During 2011 alone, PV module prices fell by almost 50%, according to BNEF.

Costs of wind turbines and wind-generated power also continue to decline. BNEF has estimated that, between 2009 and 2011, the price of wind turbines fell by 18%, reflecting fierce competition in the supply chain. During 2011, prices for onshore wind turbines declined by 5–10%.

Further Reductions Expected

Government policies are still needed in most regions of the world to address remaining barriers to renewable energy, support deployment, and help drive down costs even more. Further cost reductions are expected for all renewable technologies over the coming years as technologies, supply systems, manufacturing processes, and other factors continue to evolve and advance.

According to the IPCC SRREN, potential areas for cost reductions include:

- Improvements in foundation and turbine designs for offshore wind energy
- New and improved production of biomass feedstocks and supply systems
- Advanced biofuels processes and biorefining
- Advanced technologies and manufacturing processes for both PV and CSP
- Enhanced geothermal power systems
- A variety of possible ocean energy technologies under development

As the most mature renewable technology, hydropower has fewer opportunities for cost reductions, but there exists the potential to improve project performance and to make future projects technically feasible in a wider range of locations.

1.7. POLICY TRENDS

A number of factors are driving the rapid growth in renewable energy markets, increased investment, and reductions in costs. But government policies have played a crucial role, if not the most important role, in driving investment and accelerating the deployment of renewable energy technologies.

More Policies in more Countries

Although some governments began enacting policies to promote renewables after the first oil crisis in the early 1970s, relatively few countries had renewable energy policies in place before the early 1990s. Since then, and particularly over the past decade, policies have begun to emerge in a growing number of countries and at all levels of government.

According to the *REN21 Renewables Global Status Report* series, by 2004 at least 48 countries, including 14 developing countries, had some type of policy in place to support renewable electricity. By early 2012, the total number of countries and territories with policies to support renewable power had doubled to at least 109, up from at least 96 the previous year, and more than half of these countries were developing or "emerging" economies. Most countries with policies have more than one type of policy in place, and many policies have been amended or strengthened over time.

Existing Policies and Sectors

To date, most policies have focused on supporting renewable energy deployment in the electricity sector, but existing policies are directed to all end-use sectors, and attention to renewable transport and heating/cooling is increasing year by year. For example, in 12 countries analyzed for the IEA in 2007, the number of policies enacted related to heating (directly or indirectly) rose from 5 in 1990 to more than 55 by mid-2007. The REN21 *Renewables 2012 Global Status Report* estimates that by early 2012 about 19 countries had specific renewable heating/cooling targets in place and at least 17 countries or states had obligations to promote renewable heating.

Governments continue to enact more policies and to update and revise existing policies in light of lessons learned about design and implementation, and in response to changes in the market and/or technology advances and cost reductions. While most policy revisions have resulted in more supportive or aggressive policies, some revisions have resulted in reduced policy support. In 2010 and 2011, for instance, a number of countries reduced payments for solar PV generation under existing feed-in tariffs (FITs) in response to rapid increases in installed capacity and falling module prices, while others reduced renewable energy incentives due in part to the international economic crisis.

Regarding policies, some general observations include:

Most commonly used policies. For electricity from renewable sources, quotas and FITs are the most commonly used policies. In the heating/cooling and transportation sectors there has been a trend from fiscal incentives toward mandates.

Renewable energy targets. Increasingly, countries are also adopting political targets (or strengthening existing targets) for renewable energy for specific shares of electricity, transportation fuel, heat, or primary or final energy from renewable sources. By early 2011, at least 118 countries had such targets, up from an estimated 109 countries in 2010. While most countries have both policies and targets, some countries (such as the United States) have national policies without obligatory targets, and many (particularly developing countries) have targets but have not yet enacted policies to support them.

Local policies. The number and variety of policies and targets is also on the rise at the state/provincial and municipal levels. Several hundred city and local governments have set goals or enacted policies to support renewable energy, up to a transition toward meeting 100% of their energy demand with renewable sources.

International policies and partnerships. The number of international policies and partnerships is also on the rise. For example, in 2009 the European Union put into force a directive that established a binding regional target to source 20% of the final EU energy consumption from renewables by 2020. Another example is the Mediterranean Solar Plan, a regional agreement for research and the construction of 20 GW of renewable capacity by 2020.

Part II of this book contains a comprehensive overview and analysis of the various types of policies in use.

1.8. INTERNATIONAL PLAYERS

An increasing number of international players have entered the renewable energy space or have strengthened their engagement over the past decade, driven by a variety of factors ranging from climate change to energy security to the need for energy access among the world's poorest people, and by renewable energy's potential to address such concerns.

Appendix A contains an overview of some of the key players—international organizations and frameworks, international financial institutions and national development agencies, NGOs, and the private sector—and brief descriptions of their involvement in the renewable energy sector. These players offer knowledge, marketing power, and the potential for cooperation, partnerships, financing opportunities, and other resources that might aid and inspire policy- and key decision makers.

1.9. NEAR-TERM MARKET OUTLOOK

In response to the above market, policy, investment and other trends, renewable energy deployment is expected to continue its rapid rise in the years to come. Looking to the near future, things appear bright for several renewable energy technologies, with significant capacity of solar PV, CSP, geothermal, and other technologies in the pipeline across the globe. Projections to 2015 for wind and solar power show that rapid growth is expected to continue in coming years.

1.10. GLOBAL WIND MARKET PROJECTIONS

BTM Consult (a part of Navigant Consulting) projected in early 2011 that by 2015 global wind power capacity would reach nearly 514 GW, and that it could approach 1,000 GW by the end of 2020—generating enough electricity to meet more than 9% of global electricity demand (based on IEA projections of total demand), up from about 2% in 2010. The Global Wind Energy Council (GWEC) has projected that global capacity will reach 500 GW by the end of 2016, driven mainly by China, but with strong growth also in North America and Europe (see Figures 1.4 and 1.5). GWEC also showed, in three long-term scenarios, that global wind power capacity could reach 2,300 GW by 2030, providing up to 22% of the world's electricity needs by then.

GWEC's projections for 2011 were very close to reality. However, projections for future years do not account for other global developments in 2011, including the increased turmoil in financial markets and Japan's Fukushima nuclear accident in March 2011. So, a great deal of uncertainty surrounds the 2012–2016 numbers in Figure 1.4.

Market Forecast 2012-2016

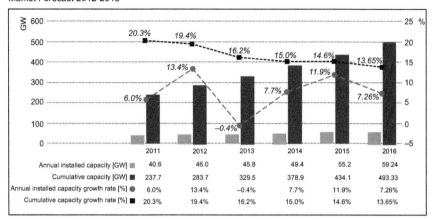

	2011	2012	2013	2014	2015	2016
Annual installed capacity [GW] ■	40.6	46.0	45.8	49.4	55.2	59.24
Cumulative capacity [GW] ■	237.7	283.7	329.5	378.9	434.1	493.33
Annual installed capacity growth rate [%] ●	6.0%	13.4%	−0.4%	7.7%	11.9%	7.26%
Cumulative capacity growth rate [%] ■	20.3%	19.4%	16.2%	15.0%	14.6%	13.65%

FIGURE 1.4 GWEC Wind Power Market Forecast, 2012–2016. *(From GWEC's Global Wind Report; Annual Update 2011)*. For color version of this figure, the reader is referred to the online version of this book.

Annual Market Forecast by Region 2012-2016

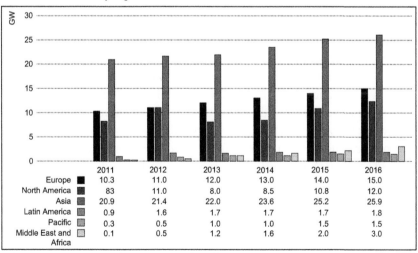

	2011	2012	2013	2014	2015	2016
Europe ■	10.3	11.0	12.0	13.0	14.0	15.0
North America ■	83	11.0	8.0	8.5	10.8	12.0
Asia ■	20.9	21.4	22.0	23.6	25.2	25.9
Latin America ■	0.9	1.6	1.7	1.7	1.7	1.8
Pacific □	0.3	0.5	1.0	1.0	1.5	1.5
Middle East and Africa □	0.1	0.5	1.2	1.6	2.0	3.0

FIGURE 1.5 GWEC Wind Power Annual Market Forecast by Region, 2012–2016. *(From GWEC's Global Wind Report; Annual Update 2011)*. For color version of this figure, the reader is referred to the online version of this book.

1.11. GLOBAL SOLAR PV MARKET PROJECTIONS

The European Photovoltaic Industry Association (EPIA) projected in early 2012 that between 208 and 343 GW of PV will be installed globally by 2016, up from about 40 GW at the end of 2010. Compared to 2011 projections, 2015 figures had to be corrected upward, indicating that the increase in reality is closer to the optimistic projections than to the "moderate" projections. The EU will likely remain the world leader during this period, but strong growth will be seen elsewhere around the world, particularly (as with wind power) in Asia (see Figures 1.6 and 1.7) EPIA's policy-driven forecast (which has proven most accurate in past years) predicts annual growth rates averaging more than 30% during this period; even the "moderate" scenario shows average annual growth rates of at least 20%. However, EPIA cautions that until PV has achieved competitiveness in all market segments, PV market growth will be highly dependent on government policies.

In a 2011 publication, *Solar Energy Perspectives,* the IEA also projects rapid growth for solar PV. According to the report, solar energy could provide one-third of the global final energy demand by 2060, but only if "all necessary policies are implemented rapidly". And even that projection could be conservative in light of recent price reductions for solar systems, which are at or approaching grid parity in some locations.

It is worth noting that a 2009 survey, produced for the German Agency for Renewable Energy, reviewed almost 50 forecasts (published between 1980 and 2007) for renewable energy growth and found that nearly all of them had underestimated the actual increase in renewable energy deployment or share of energy provided by renewable sources. The projections for Germany, Europe, and the world were made by a range of scientific and political institutions as well as industry associations, including the IEA, European Commission, German and U.S. government agencies, the European Wind Energy Association, and Greenpeace. They covered a number of specific technologies (e.g., wind power, solar PV, solar heat, geothermal heat pumps, biomass power), shares of total electricity, or of total energy. Some of the early projections failed to account for certain technologies that have become major players in recent years; the few early projections that overestimated renewables' future share assumed both stronger support policies and higher energy prices than those that materialized by the year 2000.

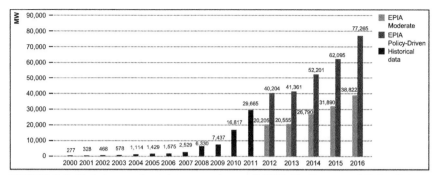

FIGURE 1.6 Global annual market scenarios until 2016 Moderate and Policy-Driven (MW) (2012–2016). *(From Global Market Outlook, May 2012).* For color version of this figure, the reader is referred to the online version of this book.

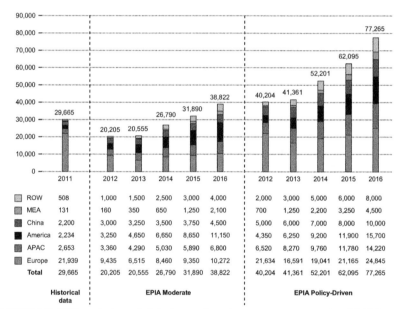

FIGURE 1.7 EPIA evolution of global annual market scenarios per region (MW) (2012–2016). *(From Global Market Outlook, May 2012).* For color version of this figure, the reader is referred to the online version of this book.

Moving forward, however, the global community cannot afford to be complacent based on past achievements. To realize continued strong growth for the future—and particularly to move beyond current projections and to achieve aggressive targets for all renewables over the next five years—stepped-up policies and political targets will be required from more countries around the world.

1.12. TO GET READY FOR THE NEXT STEP

Significant developments in recent years with regard to market and industry growth, technology advancements and cost reductions, and investments highlighted in this chapter provide evidence that renewable energy has entered the mainstream. At this stage, one is justified in asking whether renewable energy technologies are advancing quickly enough to enable the world to meet its growing demand for energy in a sustainable manner.

Chapter 2 introduces several projections and scenarios from authoritative sources, and discusses their implications for the environmental, economic, and social sustainability of the world's future energy system.

CHAPTER TWO

What Is Possible and by When?

Contents

To what extent can low–carbon energy technologies, particularly renewable energy technologies combined with energy efficiency improvements, contribute to the 2°C target? And by when? Several relevant scenarios present robust projections and provide some insight into the probability and margins of predicted future developments.

2.1. RELEVANT AND AUTHORITATIVE SCENARIOS

Scenarios are the result of modeling possible future developments in energy supply and demand systems. In general, they are highly dependent on assumptions and other inputs about a number of parameters, including energy prices, costs, climate change developments, future policies, availability of resources, and even geopolitical issues (which affect resource availability and cost). As such, scenarios are not forecasts; however, they are helpful instruments that decision makers in government and business can employ when designing strategies and specific policies. For this reason, the Implementing Agreement on Renewable Energy Technology Deployment (IEA-RETD) has monitored the role of renewable energy in global energy modeling since 2007.

This section summarizes the results of several recent energy scenarios including an overview of scenarios in the Intergovernmental Panel on Climate Change (IPCC) *Special Report on Renewable Energy Sources and Climate Change Mitigation*, several scenarios from the International Energy Agency's (IEA) *World Energy Outlook* (WEO), and the IEA's Blue Map scenario set out in the *Energy Technology Perspectives 2010* (*ETP2010*) and the IEA-RETD's *Achieving Climate and Energy Security* (ACES) scenario, as well as paths toward 100% renewables produced by the World Wildlife Fund (WWF) and by Greenpeace International with the European Renewable Energy Council (EREC).

Renewable Energy Action on Deployment
http://dx.doi.org/10.1016/B978-0-12-405519-3.00002-5

This summary of scenarios is intended to provide an overview of possible future developments within the global energy system. Except for IEA-RETD's own scenario, the scenarios included in this chapter do not necessarily reflect views of the IEA-RETD community.

Table 2.1 outlines the most important aspects of the scenarios described below.

2.2. THE SCENARIOS

IPCC Special Report on Renewable Energy Sources and Climate Change Mitigation (2011)

The Special Report on Renewable Energy Sources and Climate Change Mitigation (SRREN) reviewed 164 scenarios from the existing peer-reviewed literature (see Figure 2.1). It concluded the following:

- Baseline scenarios show a virtual "lower limit" to the supply of energy derived from renewables in 2050. Many "business as usual" scenarios show that by 2050 renewables will provide at least 100 exajoules (EJ) of primary energy, with renewables in some scenarios exceeding 250 EJ; this compares to renewable energy's primary energy production of 64 EJ in 2008 (or 12.8% of a total 500 EJ primary energy consumption).
- Assuming that the greenhouse gas (GHG) concentration in the atmosphere must be stabilized at a maximum 440 parts per million CO_2-equivalent to remain below the 2°C threshold, the median deployment of renewables by 2050 is 248 EJ/year, with a maximum of 428 EJ/year.
- More than half of the scenarios show a renewables' contribution of more than 27% of global primary energy in 2050; the scenarios with the highest renewable energy shares reach approximately 43% in 2030 and 77% in 2050. Other low-carbon contributions include nuclear energy and fossil fuels combined with carbon storage.
- Scenarios generally conclude that renewable energy growth will be widespread around the globe, although the distribution of renewable energy deployment varies substantially among regions.
- No single renewable energy source will be dominant on a global basis.

IEA World Energy Outlook

Every year the IEA produces its World Energy Outlook (WEO) scenario, an independent analysis and assessment of expected future energy developments. Because the outlook is published annually, the series provides a good insight into recent shifts in the modeling of future energy developments.

The last four WEO scenarios (published annually from 2008 through 2011) have focused primarily on developments to 2030 or 2035. With each passing year there has been a clear shift toward a more prominent role for renewable energy in the future world energy mix. This shift has emerged in response to more aggressive national and regional climate actions and targets, and is also due to increased attention in the model to policy drivers such as security of supply, an expected increase in decentralized generation of energy (replacing large central systems), and new business models. Instead of looking at energy systems in isolation, more recent scenarios have also included more calculations on the comprehensive costs and benefits of a transition (see the sections Economics of the Transition and Employment, 2.2.7.).

An overview of successive WEOs:

WEO 2008

- Focused on a reference scenario, based on business-as-usual policies and other assumptions, with high growth in demand and supply, and with large shares of fossil fuels.
- Additional scenarios modeled the likely developments under CO_2 concentration limits of 550 and 450 ppm.

WEO 2009

- The reference scenario remained the most prominent scenario. It assumed that, under business-as-usual, atmospheric CO_2 concentrations would stabilize around 2100 at almost 1000 ppm. It was slightly adapted from the 2008 version to account for new findings and the early experiences with the global financial crisis in 2008/2009.
- The WEO 2009 skipped the 550 ppm scenario to focus on the 450 scenario, which showed global emissions peaking by 2020, due primarily to emissions reductions in the energy sector through energy-efficiency improvements and increased deployment of renewable energy technologies (see Figure 2.2).
- The renewables contribution by 2030 in the 450 ppm scenario was estimated at about 22%, excluding large hydro (which accounted for another 4%). Biomass represented the largest share of this contribution, or ~16% of all global (primary) energy demand (see Figure 2.3).

WEO 2010

- In WEO 2010, the IEA put a "New Policies" scenario between the Current Policies scenario (the follow-up of the reference scenario) and the 450 scenario, focusing the outlook on 2035. The New Policies scenario was based on the new global context of the 2009 Copenhagen Accord,

TABLE 2.1 Overview of the scenarios in this chapter

Global scenario	Year of publication	Organization	Time frame	CO_{2eq} emissions reduction target	Share of renewable energy	Costs and benefits	Additional description
SRREN	2011	IPCC	Ranges from 2030 to 2050 and beyond	Baselines 2050: 40 to 85 Gt CO_2 per year; scenarios: 0 to 55 Gt (present: 30 Gt/year)	More than half of 2050 scenarios project >173 EJ/year, up to >400 EJ/year	Four illustrative scenarios: annual costs <1% of GDP, through at least 2030	Review of 164 peer-reviewed scenarios
WEO 2011	2010	IEA	2035	450 scenario: 49% emissions reduction compared to Current Policies scenario (which shows reductions about 30% below present levels)	2035: 45% of electricity generation; 21% of heat; biofuels 14% of transport fuel	Cost: USD 18 trillion additional for 2010–2035 relative to Current Policies; benefit: fuel cost savings of USD 17.1 trillion (calculated over the lifetime of implemented technologies)	Considers aggressive implementation of Copenhagen Accord; stabilization of atmospheric CO_2 concentration at 450 ppm
WEO 2011	2011	IEA	2035	450 scenario: 41% emissions reduction compared to the New Policies scenario (or around 30% reduction compared to present levels)	2035: renewables achieve 27% share overall; 46% of electricity generation; biofuels 13% of transport fuel; heat not specified	Cost: USD 180 billion/year additional compared to New Policies scenario from 2011 to 20; 1.1 trillion/year additional during 2021–2035; cumulative cost during 2011–2035: USD 15.2 trillion	Emphasizes that locked-in investments narrow window of opportunity to achieve 2°C goal

Scenario	Year	Organization	Target year	Emissions target	Renewable share	Costs / benefits	Notes
ETP 2010	2010	IEA	2050	50% emissions reduction relative to 2010 levels	48% of electricity generation; others not specified	Costs: USD 46 trillion (until 2050) above baseline; benefit: fuel cost savings of USD 66 trillion (over lifetime of implemented technologies)	The central Blue Map scenario in *ETP* is based on 450 scenario from WEO 2010 (2035), extended to 2050
ACES	2010	IEA–RETD	2060	Limit atmospheric concentration to 400 ppm by 2100	60% of final energy use	Costs: direct net costs <1% of global GDP, or possibly negative relative to baselines over period until 2060	Objective was to determine what realistic measures exist to stay below 400 ppm of CO_2-equivalent
The Energy Report	2011	WWF/ Ecofys	2050	Almost 100% (driven by renewable energy target)	95% of final energy use	Net costs: 1–2% of GDP annually until 2035; after 2035 tipping point, a growing positive balance, meaning benefits will exceed costs (relative economic benefits of +2% by 2050, compared with business as usual)	Strong energy and materials efficiency; calculations for >95% renewables by 2050 are bottom-up per sector and per technology
Energy [R] evolution	2010 (update)	Green- peace/ EREC	2050	>80% below current emissions levels	80+% of final energy use	Cost: maximum USD 31 billion/year by 2020 compared to the reference scenario; benefit: USD 2.7 trillion annual savings in fuel costs by 2050	Strong focus on a "green jobs" revolution; 2015: 4.5 million additional jobs in the energy sector; 2030: 3.2 million additional jobs

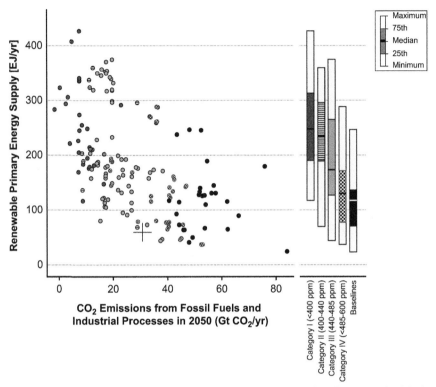

FIGURE 2.1 Shows 164 scenarios, summarized in one graph by the IPCC. The black dots at the right side represent business-as-usual scenarios, while four categories of scenarios show a wide range of corresponding renewable contribution (*y*-axis, in EJ) and GHG emissions (*x*-axis). The gray crossed lines show the relationship in 2007. *(From: IPCC 2011: Special Report on Renewable Energy Sources and Climate Change Mitigation, Summary for Policymakers, Figure SPM.9. Cambridge University Press).* For color version of this figure, the reader is referred to the online version of this book.

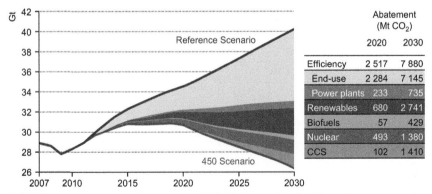

FIGURE 2.2 The 450 (ppm) scenario in WEO 2009 compared to the reference scenario. The contributions of different sources are represented on the right. For color version of this figure, the reader is referred to the online version of this book.

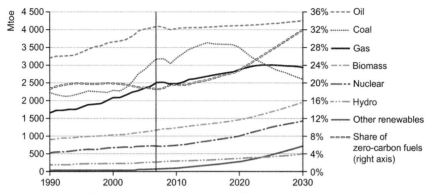

FIGURE 2.3 Deployment of contributions by different energy sources until 2030. The dotted black line includes nuclear energy. For color version of this figure, the reader is referred to the online version of this book.

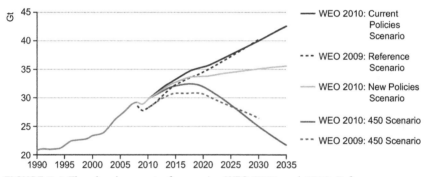

FIGURE 2.4 The development of scenarios WEO 2009 and 2010. Reference scenarios converge toward 2030/2035, but the WEO 2010 Current Policies scenario shows increased CO₂ emissions relative to the WEO 2009 version. The 2010 New Policies scenario includes new country pledges and the 450 scenario accounts for new insights. For color version of this figure, the reader is referred to the online version of this book.

including all national pledges and announced policy intentions. In the update the IEA again increased the emphasis on possible low-carbon routes. The 450 ppm scenario identifies a 22% share of modern renewable sources (excluding traditional biomass) in primary energy demand by 2035 (see Figures 2.4 and 2.5).

WEO 2011

- WEO 2011 replaced the Current Policies scenario with the New Policies scenario as a baseline. Much emphasis was placed on the 450 scenario, which provided a 50/50 chance of global warming not exceeding the 2°C threshold (Figure 2.6).

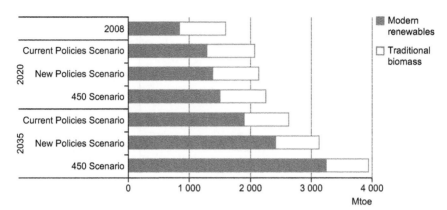

FIGURE 2.5 World primary renewable energy supply by scenario (WEO 2010).

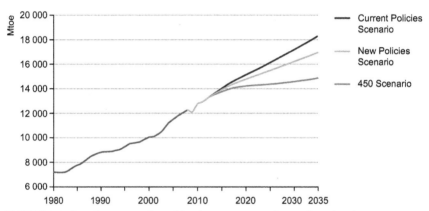

FIGURE 2.6 Development of world primary energy demand under three scenarios (WEO 2011). For color version of this figure, the reader is referred to the online version of this book.

- The report emphasizes that installations currently operating or planned already account for ~80% of all energy-related CO_2 emissions allowed up to 2035 if warming is to remain below the 2°C threshold.
- For every USD1 of investment avoided in the power sector before 2020 an additional USD 4.30 would need to be spent after 2020 to compensate for the increased emissions.
- In the 450 scenario, the amount of renewable energy deployed in 2035 will be slightly higher than that in the WEO 2010 scenario. The renewables' share will be 27% of total primary energy demand overall, with renewables providing 46% of electricity generation and biofuels accounting for 13% of transport fuel (see Figure 2.7).

- In the power sector, renewable energy technologies, led by hydropower and wind power, will account for half of the new capacity installed over the period up to 2035 to meet growing demand.
- Renewable shares in the power sector will be larger if carbon capture or nuclear energy is not deployed further. However, costs will rise (Figure 2.8).
- Accommodating more electricity from renewable sources will require additional investment in transmission networks, amounting to an average 10% of total transmission investment worldwide and 25% in the European Union.

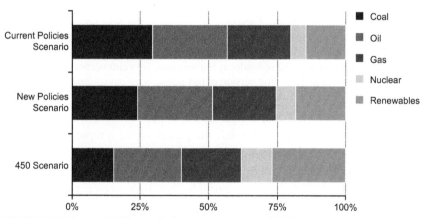

FIGURE 2.7 Shares of different fuel in three scenarios (WEO 2011). For color version of this figure, the reader is referred to the online version of this book.

Pathways in *Energy Technology Perspectives 2010*

The IEA publication *Energy Technology Perspectives 2010* (*ETP2010*) presents its options for renewable energy technologies within the context of expected economic developments to 2050. This report goes beyond the IEA baseline scenario in the WEO by presenting a "Blue Map" scenario (and some variations), which focuses on a 50% reduction in annual GHG emissions by 2050.

The Blue Map scenario assumes that by 2050 the Organization for Economic Cooperation and Development (OECD) countries will have reduced their emissions by 70–80% and non-OECD countries by 30% (compared to 2007). It shows an increase of global primary energy demand by 32%, compared with the 84% increase in the baseline scenario. Electrification in

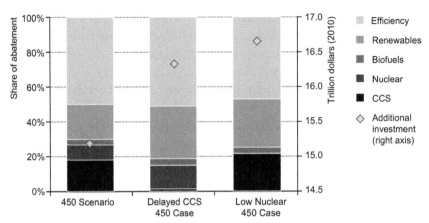

FIGURE 2.8 Cumulative share of abatement relative to the New Policies scenario in the 450 scenario, with sensitivity analysis regarding delayed development of other low-carbon options, such as CCS or nuclear energy. Even more renewable energy will be required than in the basic 450 scenario, at higher costs. For color version of this figure, the reader is referred to the online version of this book.

end-use sectors such as transport and in buildings will facilitate a substantial reduction in emissions, provided that the power sector is rapidly decarbonized. The *ETP2010* also presents options, rather than forecasts, and roadmaps for possible future development.

The aim of *ETP2010* is to provide "an IEA perspective on how low-carbon energy technologies can contribute to deep CO_2 emissions reduction targets. Using a techno-economic approach that assesses costs and benefits, the book examines least-cost pathways for meeting energy policy goals while also proposing measures to overcome technical and policy barriers." *ETP2010* also demonstrates that low-carbon technologies can be powerful tools for enhancing energy security and economic development.

The *ETP2010* concludes that the next decade is critical: "If emissions do not peak by around 2020 and decline steadily thereafter, achieving the needed 50% reduction by 2050 will become much more costly." Meanwhile, it notes that "an energy revolution is within reach. Achieving it will stretch the capacities of all stakeholders." Most importantly, a broad portfolio of existing and new technologies will be needed alongside significant improvements in energy efficiency. The *ETP2010* lays out priority areas and mechanisms that can help to deliver these changes. It states that governments will need to intervene on an unprecedented level to encourage investors and businesses to take the lead (see Figure 2.9).

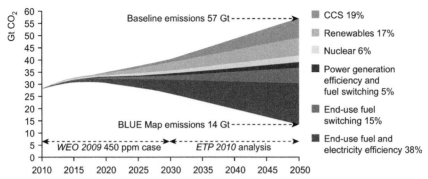

FIGURE 2.9 GHG emissions according to the WEO 2009 scenario (until 2030), extended to 2050 by the Blue Map scenario. Blue Map shows a 50% global emissions reduction by 2050 compared to 2010, and even of about 75% compared to the baseline.

In addition, the *ETP2010* made a rough estimate of the amount of investment required to achieve the Blue Map scenario. It projects that ~17% additional (USD 46 trillion) investment, beyond that needed to realize the baseline scenario (an accumulated USD 270 trillion, excluding upstream investments), will be required for the 50% emissions reduction scenario. However, the *ETP2010* concludes that these extra investments will be more than offset by fuel savings, estimated at an accumulated USD 112 trillion; further "co-impacts," such as an increased security of supply or on environmental improvements, could provide additional economic benefits/offsets.

Even excluding impacts for an increased security of supply or environmental improvements, the extra investments for renewable energy are more than offset by fuel savings.

Pathways in *Energy Technology Perspectives 2012*

The IEA publication *Energy Technology Perspectives 2012 (ETP2012)* states that "a technological transformation of the energy system is still possible, despite current trends." *ETP2012* emphasizes that some technologies that hold large potentials are failing to meet the necessary transition to a low-carbon economy. With that remark it especially targets energy efficiency and carbon capture and storage (CCS). Some more mature renewable energy technologies (hydro, biomass, onshore wind, solar photovoltaic; PV) are making good progress, while offshore wind and concentrated solar power are lagging behind.

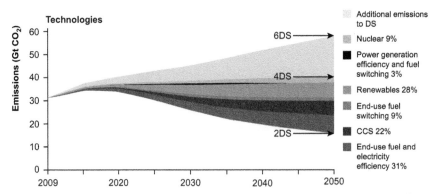

FIGURE 2.10 Emissions trajectory in the *ETP2012* 2DS, including the contributions from different technologies. Percentages are relative to the 4DS, which is largely a business-as-usual scenario. *(From ETP2012, Figure 9).* For interpretation of the references to color in this figure legend, the reader is referred to the online version of this book.

While energy efficiency is regarded as a key, "low-carbon electricity is at the core of a sustainable energy system." *ETP2012* explores three energy scenarios to 2050, resulting in a global temperature rise of, respectively 2, 4, and 6°C. The focus is on the 2° scenario (2DS), which shows an emissions trajectory similar to the IEA WEO 450 scenario to 2035 (see Figure 2.10). The 4° scenario (4DS) is a reference scenario that resembles the WEO New Policies scenario and is regarded as a plausible scenario, "but it is clear that governments must play a lead role." The 6° scenario (6DS) is a reference scenario without any new policies.

The figures slightly differ from the *ETP2010* report. *ETP2012* confirms that future savings from an ambition scenario like 2DS more than compensate for the up-front investment costs. A net economic surplus at the global level would exist, due to the value of fuel savings, estimated at USD 100 trillion between 2010 and 2050. Even at high discount rates and without taking into account avoiding climate change damage costs, a net saving of trillions will occur. Meanwhile, *ETP2012* also confirms that there will not be only winners. "Some regions and sectors will undoubtedly come out worse from an economic standpoint in the 2DS, but the overall picture looks surprisingly good."

The IEA-RETD ACES Scenario

The IEA-RETD, in collaboration with the IEA Implementing Agreement Energy Technology Systems Analysis Programme (ETSAP), developed the ACES scenario. It is based on a techno-economic, bottom-up model, and places constraints on the global trade of energy commodities to reflect the rising importance of energy security. The ACES scenario also accounts for specific characteristics of renewable energy resources and technologies, such as variability, and incorporates the impacts of concepts of so-called "smart grids," that can enable efficient balancing of demand and supply. In order to evaluate the results of the IEA-RETD ACES scenario, a reference scenario was developed that assumes no future policy changes.

ACES was modeled using the authoritative ETSAP-TIAM3 economic optimization model, developed and regularly updated within IEA ETSAP. The economic model balances supply and demand for energy services by maximizing total economic surplus for users and producers within certain constraints such as emissions limits. The model covers all relevant energy supply and demand technologies divided over 16 world regions, based on IEA numbers. The model and its derivatives have been used for numerous modeling studies and scenarios, including 10 of the 164 scenarios reviewed in the IPCC SRREN.

The main target in ACES was to stabilize global GHG concentrations at 400 ppm CO_2-equivalent within this century, in line with international objectives to prevent severe and irreversible climate change. However, this target was not achieved under the TIAM3 bottom-up model; instead, concentrations peak in 2035 (at about 490 ppm) and then gradually decline to 420 ppm by 2100.

Because the ACES concentration peak does not occur until 2035, remaining below the targeted 2°C threshold can be achieved only by realizing large "net negative emissions" from 2060 onward. This means that significant amounts of CO_2 must be removed from the atmosphere, for example, by reforestation (uptake by trees), and the use of biomass for power combined with carbon capture and sequestration technologies.

The emission reduction is caused primarily by three important developments within the energy sector: a higher efficiency in energy conversion, a lower demand, and a rapid decarbonization of the energy supply via renewable energy sources and capture and storage of CO_2 with fossil fuels (see Figure 2.11).

In the IEA-RETD ACES scenario, the electricity sector undergoes far greater transformation than any other sector and is fully decarbonized by 2030 (see Figure 2.11). This results from a rapid reduction in the use of fossil fuels for power generation, except where this usage is also accompanied by CCS, with renewables becoming the largest contributor before 2030. By 2030, renewables will contribute 61% of global electricity generation

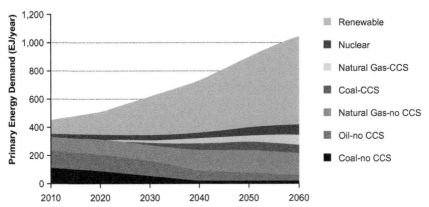

FIGURE 2.11 The IEA-RETD ACES scenario shows possible developments toward 2060 that are considered to be realistic. The largest contribution to energy demand comes from renewable energy technologies with significant contributions also from fossil fuels with CCS and natural gas, while coal and oil without CCS gradually fall toward zero. For color version of this figure, the reader is referred to the online version of this book.

and, between 2030 and 2040, renewable energy will exceed the 50% share in primary energy supply. All renewable energy technologies grow significantly in this scenario, although wind and solar PV see the largest relative growth.

Compared to the reference scenario, about 35% less energy is traded across borders in the IEA-RETD ACES scenario. The preference for the consumption of domestic (renewable) resources increases the security of supply for individual nations considerably.

The ACES study assesses the costs and investments that will be required under the scenario. The total direct energy system costs for ACES exceed the reference scenario costs by USD 14.3 billion over the next 50 years. The accumulated additional cost is equivalent to about 1% of the accumulated world Gross Domestic Product (GDP) during this period, which is in the same order of magnitude as many other 2DSs.

Although the ACES study did not calculate economic benefits of this scenario, it cites other reports that have determined that total benefits will most likely exceed costs. This study concludes that "Aggressive climate change mitigation has minimal incremental costs and may even be economically superior to inaction." This statement applies only to direct energy cost savings; indirect benefits such as reduced costs associated with climate change and reduced environmental damages would be additional.

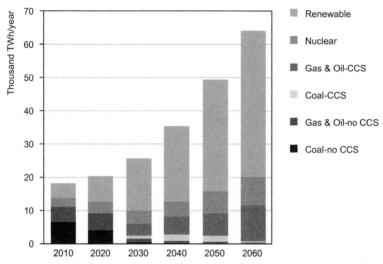

FIGURE 2.12 The ACES scenario describes a rapid reduction in the share of nuclear and fossil energy in the power sector over the next two decades. Overall electricity generation will continue to grow exponentially during the twenty-first century. For color version of this figure, the reader is referred to the online version of this book.

Paths approaching 100% renewables

In recent years, some scenarios have been developed to examine pathways for achieving (almost) 100% renewable energy by 2050. They go beyond the scenarios described above by maximizing the input of energy efficiency and renewable energy technologies. This implies future adoption and implementation of more and stronger policies that maximize improvements in energy efficiency and accelerate deployment of renewables, leading to a major shift in investment from fossil fuels to renewable energy. These paths are included here as examples of potential far-reaching changes.

The Energy Report of WWF (based on a scenario by Ecofys) and *the Energy [R]evolution* of Greenpeace and the EREC are examples of these (near-) 100% scenarios.

The Energy Report

The Energy Report (2011) uses a bottom–up approach with the aim of answering the question: "Is a fully sustainable global energy system possible by 2050?" The scenario first forecasts a future level of energy–consuming activities (e.g., tonnes of steel used), based on a growing population and a projected tripling of the world's economy by 2050. It assumes that these services would be delivered with a minimum of energy and material use. Second, it ranks all renewable sources available to provide the energy for these activities. The sun, wind, water (ocean and hydro), and geothermal heat are prioritized, while (sustainable) bio-energy is also deployed. By 2050, only some particular manufacturing processes (e.g., steel, cement) will still need fossil fuels for their specific properties that will not yet be substitutable.

An important condition in the scenario is the development toward a sustainable standard of living for all people around the world, based on increasing equity in living standards and corresponding per capita CO_2 emissions. Substantial up-front financing will be required, but economic benefits are expected to outweigh costs over the long term.

Energy [R]evolution

The Energy [R]evolution scenarios are developed by Greenpeace International and the EREC, with modeling done by the German Aerospace Center; the third edition was published in 2010. These scenarios sketch pathways to an almost fully renewable energy system in which, by 2050, renewables account for 95% of electricity generation and 80% of total global (final) energy demand.

Energy [R]evolution: A Sustainable World Energy Outlook combines a reduction in carbon emissions alongside economic growth by replacing fossil fuels with renewable energy and significant improvements in energy efficiency. Much emphasis is placed on the creation of millions of new green jobs. As in The Energy Report and many other scenarios, electrification of the transportation and heating and cooling sectors are essential elements in the transition to a sustainable energy supply.

Delaying action increases the costs

Several scenarios show that global CO_2 emissions must peak before 2020 if the world is to avoid exceeding the 2°C threshold. The longer emissions are permitted to rise, the greater the associated costs will be, for two reasons: the scale of damages and subsequent required adaptation will be larger (along with other environmental, social, and security costs associated with fossil fuels) and low-carbon technologies will need to be deployed more rapidly, which is likely to result in a higher average cost. This implies that starting the needed transition sooner rather than later, by significantly accelerating the deployment (and share) of renewable energy, will reduce the costs associated with climate change and its mitigation. Looking at the energy transition in this way provides a favorable outlook regarding the costs and benefits of renewable energy, including the potential employment benefits of renewables deployment.

Economics of the transition

The estimated economic costs of a transition to a low-carbon economy that is based primarily on renewable energy are relatively well understood and quantified. These costs concern primarily the incremental costs for investments in new technologies and the corresponding infrastructure relative to the investments that would be required under business-as-usual development (with large contributions of fossil fuels). Many models quantify these marginal costs, taking into account learning effects, gradual cost reductions, and required investments in R&D.

In contrast, the economic benefits of such a transition have not been fully examined and are difficult to quantify. However, many studies provide evidence that the costs of inaction on climate change are significantly greater than the costs of mitigation; that aggressive climate change mitigation is, in fact, economically sound; and that renewable energy (combined with energy efficiency improvements) is a critical component of a least-cost strategy to achieve this mitigation.

The IEA *World Energy Outlook 2011* concluded that: "Delaying action is a false economy: for every USD 1 of investment avoided in the power sector before 2020 an additional USD 4.30 would need to be spent after 2020 to compensate for the increased emissions." Further, a recent IEA-RETD report, *Climate Change Adaptation, Damages and Fossil Fuel Dependence (2011)*, found that the costs of inaction are large and appear to be significantly greater than the costs associated with climate change mitigation.

In broad terms, the IEA-RETD report concluded that the costs of climate adaptation and damages in a scenario with minimal mitigation are likely to be in the range of several hundred billions of dollars annually by 2030. This conclusion is based on current studies regarding adaptation that estimate costs in the range of USD 50–200 billion per year by 2030, and additional analysis of such estimates, which suggests that these costs might be underestimated by a factor of three. Residual costs associated with climate change—damages that cannot be alleviated through adaptation—during the same time frame are also estimated at several hundred billion dollars annually.

The IEA's 2009 WEO 450 scenario estimates the incremental costs of associated mitigation efforts to be USD 10.5 trillion from 2010 to 2030, with USD 8.3 trillion devoted to transport, buildings, and industry, and the remaining USD 2.2 trillion going to power generation and biofuels supply. Notably, USD 4.75 trillion goes to transportation in general, and one-third of the total incremental investment (USD 3.4 trillion) is marked for light-duty vehicles.

While the study did not estimate health-related benefits or reduction in other external costs, it did quantify the expected reduction in energy expenditures. Energy bills in transport, buildings, and industry were expected to decline by USD 8.6 trillion from 2010 to 2030 under the 450 scenario; this implies a net savings of USD 300 billion based on fuel savings alone compared to the incremental investment costs needed in these sectors of USD 8.3 trillion. Moreover, the undiscounted cumulative fuel savings over the lifetime of these investments, beyond 2030, is estimated at USD 17 billion. This underscores the fact that shifting the economy to low-carbon sources has long-term economic consequences and that the cost/benefit ratio is very favorable based on fuel savings alone.

Comparing the 20-year USD 10.5 trillion of needed incremental investment under the 450 scenario with the associated fuel savings from USD 8.6 trillion from transport, buildings, and industry (and not the power sector), implies a net cost of USD 1.9 trillion over the study period. However, this does not include other economic, health, environmental, and security benefits, nor does it quantify the long-term benefits of a shift in the economy beyond 2030. The implication is that the total net present value of benefits under the 450 scenario far exceed the economic costs.

Other estimates concur that total fossil fuel dependence costs (beyond fuel savings alone) are at least as large as climate mitigation costs. For the United States alone, the mentioned IEA-RETD study estimated total fossil fuel dependence costs at USD 450–900 billion annually under current economic conditions.

In addition to reducing fossil fuel expenditures, relative economic savings will also result from other benefits of climate change mitigation, such as cleaner air and water, and associated health improvements. Although there are no known studies that examine global environmental and health costs (beyond climate change) associated with fossil fuels, there are national and regional studies that provide a sense of the scale of such costs. According to the European Environmental Agency, the costs of health and environmental damage in Europe of pollution (including CO_2) amounted to a total of more than USD 130–210 billion (€100–160 billion) in 2009, of which two-thirds is attributed to power plants. A 2011 study by the Harvard Medical School estimated that the "hidden costs" of coal burning in the United States totaled one-third to more than one-half a trillion U.S. dollars annually, effectively doubling to tripling the per kilowatt-hour cost of electricity from coal. Costs in the Harvard study cover economic (including government subsidies), social (including USD 140–242 million/year in public health effects), and environmental impacts.

Employment

In addition to the economic benefits discussed above, several studies reveal that renewables currently create at least about as many or even more jobs per unit of energy output than do fossil fuels. Many of these jobs are located in rural areas, contributing to rural economic development. Particularly in the solar and wind sectors, jobs are primarily medium- to high-skilled positions.

A 2008 United Nations Environment Programme study estimated that renewable energy then accounted for about 2.3 million jobs worldwide. The REN21 *Renewables 2011 Global Status Report* estimated that total jobs exceeded 3.5 million by the end of 2010, with more than 1.5 million of these in the biofuels industry. It notes, however, that there are significant uncertainties surrounding these numbers that are related to issues such as accounting methods, direct versus indirect jobs, industry definition and scope, and displaced jobs from other industries (e.g., net vs. gross job creation).

In many countries, including Germany, India, Nepal, and the United States, job creation has been one of the key drivers for government policies that support renewable energy deployment. Some U.S. studies suggest that every billion dollars spent on "green measures," including renewable energy deployment, creates 20,000 to 33,000 new jobs.

Additional jobs are expected to be created along with the growth of renewable energy markets and industries, although jobs are not expected to grow at the same rate due to increased economies of scale in installation services and increasing automation in manufacturing processes.

2.3. TO GET READY FOR THE NEXT STEP

The scenarios highlighted in this chapter demonstrate that a dramatic acceleration of renewable energy deployment, leading to a major energy transition by mid-century, is realistic from both technical and economic perspectives. This acceleration should be established hand in hand with speeding up the efficiency of generation and use of energy and building new systems. If such scenarios show that a major transition is possible, why are policies crucial to bringing about the needed changes? Chapter 3 examines the barriers and challenges that renewables must still overcome to achieve the scale required.

CHAPTER THREE

Drivers and Barriers

Contents

Perspectives on renewable energy and its potentials and benefits have changed along with changing landscapes within and outside of the energy sector, including growing environmental awareness, economic crises, short-term or longer term concerns about energy security, and other issues. This chapter describes the benefits associated with renewable energy and drivers for support policies, as well as the barriers, challenges, and misperceptions that continue to act as hurdles to future deployment. Knowledge of these drivers and constraints can provide insight in the design of effective and efficient policies and measures for accelerating the advancement of renewable energy.

3.1. DRIVERS OF RENEWABLE ENERGY

Numerous benefits associated with renewable energy are driving government support policies and increased investment around the world. They include the following:

Environment. Most renewable technologies have low life cycle green house gases (GHG) emissions and relatively small impacts on biodiversity, air and water quality, and human health compared with fossil fuels and nuclear power. Renewable energy plays an integral part in government strategies to reduce carbon dioxide emissions at the national, state/provincial, and local levels.

Renewable Energy Action on Deployment
http://dx.doi.org/10.1016/B978-0-12-405519-3.00003-7

Energy security. Renewable resources exist in every country and because they are widely accessible the energy resources themselves are less vulnerable than fossil fuels to import–export issues and geopolitical problems. Renewables can reduce dependence on imported fuels. They can also help to improve the national balance of trade in many countries, diversify supply, and reduce vulnerability to price fluctuations.

Energy access. Renewables can enhance access to reliable and affordable energy services and, in many cases, offer the lowest cost options for access to energy.

Economic development. Renewables can help drive economic development, including in rural and remote areas, creating jobs and new industries. Most renewables have zero fuel costs, so operating costs are generally low, stable and predictable.

The scale and scope of benefits will vary across regions and from one technology to the next.

3.2. BARRIERS TO RENEWABLE ENERGY

There exist many examples of new technologies or systems that entered the mainstream merely on the basis of market mechanisms and end-users' demand. Cars, computers, mobile phones, and DVD players are obvious examples of innovative technologies that autonomously created large markets, infrastructure, and entire industries. In many cases, significant market demand developed because new technologies offered cheaper or more convenient alternatives to existing technologies; in other cases, innovative technologies provided entirely new and desirable services that were previous unimaginable.

In the case of renewable energy, there are a number of barriers that have impeded autonomous entry into the mainstream, and they have been explained and categorized in numerous ways in a variety of reports over the years. The Intergovernmental Panel on Climate Change (IPCC) *Special Report on Renewable Energy Sources and Climate Change Mitigation* (SRREN), in a review of the literature on this topic, presents barriers in five distinct categories: market failures, economic barriers, information and awareness barriers, sociocultural barriers, and institutional and policy barriers. Briefly, they can be described as follows:

Market failures. These include subsidies for fossil fuels or failure to internalize external costs (such as environmental or health damage costs, or pollution and climate change mitigation and adaptation costs) and benefits associated with energy production and use (see Chapter 2).

In other words, the energy market is not a free market, but it is influenced by a variety of subsidies (predominantly, both historically and currently, for fossil fuels and nuclear power rather than renewables and efficiency), vast infrastructure built to support the dominant fuels and energy technologies, and the failure to internalize a range of external costs that are paid by society as a whole, and, in some cases, by people across boundaries of space (e.g., other communities or countries) and time (e.g., future generations). Thus, renewables do not compete on a level playing field.

Economic barriers. Most renewable energy technologies face high up-front costs, which might increase financial risks experienced on the financial markets. For example, as long as renewable energy projects are considered to be risky investments, private capital markets will require higher returns, increasing costs of projects.

Information and awareness barriers. These include lack of data about natural resources, lack of skilled labor, and lack of public and institutional awareness. For example, a lack of information among professionals and the general public about the true costs of renewable energy can result in misperceptions about renewables. Such misperceptions and misunderstandings require strategies to effectively disseminate and communicate information about life cycle costs, renewable resource potentials, the current status of renewable technologies, and existing government policies, as well as programs to develop skills and the needed workforce.

Sociocultural barriers. Societal and personal values and norms can affect perception and acceptance of technologies and specific projects, and they can be slow to change. Further, most end-consumers are not interested in the origin of their energy supply; they are concerned only that their lights turn on, that their food is cold or hot, and that they have mobility when needed. Even those people who are interested in the source of supply do not always have access to such information.

Institutional and policy barriers. Many types of barriers for new technologies originate from the inertia in systems, industry, and decision making. In energy policies, this inertia is represented and maintained by existing government subsidies and regulations that evolved to support large and centralized energy systems that are based on fossil fuels and nuclear power. Permitting and approval procedures, spatial planning and integration of renewables in systems and markets are within this category.

Moreover, renewable energy, unlike new IT technologies, must compete with existing technologies that provide similar (energy) services. Fossil fuels and nuclear power have brought wealth to large parts of the world, and large industries have been built around these energy sources. Incumbent industries have played, and continue to play, a decisive role in the energy market as well as government policy and infrastructure development to date.

In addition to the SRREN, two recent International Energy Agency-Renewable Energy Technology Deployment (IEA-RETD) reports review the various barriers to renewables. Table 3.1 draws from the report Barriers, Challenges, and Opportunities (2006), summarizing several specific barriers, along with potential actions and opportunities for addressing them and the relevant stakeholders to engage in these processes.

A more recent IEA-RETD analysis (2011) lists what it concludes are the 31 most important barriers to renewable energy deployment, divided into six categories, including: policies and regulations, infrastructure, perceptions, resources, technology, and government and industry inertia (see Table 3.2).

TABLE 3.1 Summary of barriers, examples of opportunities, and stakeholders

Barriers	Opportunities	Important stakeholders
There is no level playing field for renewable energy technologies	International cooperation on: Phasing out subsidies for conventional technologies Good practice for subsides Internalisation of externalities	National governments and international forums for cooperation (UN, EU, IEA, G8)
The incentives for governments and private companies to support renewable energy development are insufficient	International agreements committing governments to demonstrate and deploy renewable energy technologies Multilateral funds for RE deployment and demonstration - partnerships with the private sector	National, regional and local authorities, international formus for cooperation (IEA, G8) the RE industry, international financing institutions
Financing is unreasonably costly for renewable energy technologies	Favourable loans for renewable energy projects through national or international institutions. Promote long-term power purchase agreements between consumers and RE generators Initiatives to stimulate carbon financing of RE projects. Training and education of financiers	International/national inancing institutions, national governments, CDM executive board, JI advisory committee, Asia-Pacific Partnership on Clean Development and Climate, energy producers, private financing institutions

Barriers	Opportunities	Important stakeholders
Technology standards are lacking for(some) renewable energy technologies and fuels	Develop standards for RE technologies, components and fuels Develop test facilities for renewable energy technologies	National and international standardisation organisations, the RE industry, industry associations, international trade organisations (WTO, NAFTA etc.)
Import tariffs and technical barriers impede trade in renewables	International cooperation on removing duties and technical barriers to trade in RE products	International and regional trade organisations (WTO, NAFTA etc.), international forums for cooperation (IEA, G8), the RE industry
Permits for new renewable energy plants are difficult to obtain	Cooperation on best practive between national and local authorities from different countries Developing internationally harmonized standard forms and requirements (to help RE project developers working internationally)	National and local authorities, the RE industry needs to adapt its products to meet requirements of authorities
Energy markets are not prepared for renewable energy	International cooperation on: best practive for grid connection/access removing market imperfections in relation to RE new interconnectors integration of intermittent RE sources promoting demand response in energy markets	National authorities, transmission system operators, distributions system operators, energy regulators, energy traders
Renewable energy skills and awareness are insufficient	Information and education aon all educational levels at national level or through international programmes Twinning between authorities and TSOs from countries with different experience and between operational personal from different countries International in-service training programmes National and international awareness compaigns.	National and local authorities, NGOs, consumers,international forums for cooperation (IEA, G8), universities and technical colleges, operational personel

TABLE 3.2 Early 2011, an IEA-RETD publication made an inventory of 31 barriers, divided among six main categories

Policies and Regulations	Perceptions	Technology
• Fossil fuel prices are too low and are not made to pay for the damages they cause • RE incentives are inconsistant and change over time • RE polices are not strong enough • RE policies are patchwork and are too complex/overtapping • Tariffs and other trade policies impede trade in RE, make RE more expensive, or stifle competition • Siting is complex and time consuming • Regulations for various sectors have negetive implications for RE	• The public, industry and policymakers all lack sufficient and accurate information about RE • Not all RE benefits are well quantified and monetized. • NIMBY • The urgency of climate change is not fully understood • Communication strategies that have been used to date have been ineffective • There is a lack of alignment among environmental groups and RE advocacy/industry groups.	• RE CAPEX is too high • RE operating costs are too high/efficiencies are to low • Not all RE technologies are "ready" • Enabling technologies are not ready • The pace of RE technology developments is too slow

Infrastructure	Resources	Government & Industry Inertia
• Grid intergration of large amounts of intermittent RE will be challenging • The electric transmission system is not designed for large amounts of remote resource • The electric distribution system is not designed to integrate large amount of distributed resources. • The workforce skills to not match the needs of the RE industry • Renewable transportation fuels are not fully fungible with existing petroleum-based fuels	• RE resources are often remote • RE resource use can lead to unintended address environmental impacts • RE resources are not always available and can result in intermittent energy output • RE resource potential is not understood well energy and quantified	• The energy system has significant inertia • Big institutions have vested interests in maintaining the status quo • The rate and scale of investment in RE is too slow/low • There is a lack of appropriate/innovative business models to drive significant growth • The RE supply chain is immature and inefficient • Political systems tend to favor incumbent industries with strong lobbies, making it difficult to implement blood policy changes.

3.3. ADDRESSING PERCEPTION AND INERTIA

Recent IEA-RETD work concludes that two barriers in particular must be addressed in order to advance renewable energy deployment at the pace and breadth required. These barriers are the misperception of renewable energy technologies (including public acceptance) and inertia in policies and industry. Both barriers call for targeted policy action in the next few years.

Overcoming these two barriers will be a "game changer." It can be expected that wherever policy makers tackle these barriers—by gathering and providing accurate information about renewables and their benefits, and collaborating with key stakeholders to overcome inertia and broaden support for renewables—a groundbreaking transformation or a step change (rather than incremental change) in the energy system will occur. Improving perception and removing inertia can create broad stakeholder support for renewables, thus paving the way to implementation of strong and consistent policies and regulations.

3.4. EVOLVING WORLD STAGE: THE CURRENT CONTEXT

Some barriers, constraints, and drivers are time dependent. General political debates, the state of climate politics, and national or global economic and political developments can all influence the dynamics of government policies and the environment in which renewable energy finds itself. Meanwhile, investors call for stability in policies and measures in order to finance new manufacturing capacity, infrastructure, and projects.

Over the past two years, a number of developments have transpired that have greatly influenced government policies and affected renewable energy barriers, drivers, and opportunities, and thus have affected the deployment of renewables in positive or negative directions. The most important of these include:

Global and regional financial crisis. Three years after the 2008 financial crisis, further challenges—including debt crises across Europe and threats to the survival of the Euro, as well as the downgrading of the U.S. credit rating—have re-ignited nervous activity on trading floors and in financial markets. The consequences for investments in renewable energy projects are not yet fully understood. If funds from banks and other private investors are unavailable or more costly (requiring high rates of return), new investments in renewable energy will slow. On the other hand, renewable energy projects may gain a competitive edge in financial markets by offering more security to investors than do other (financial) markets or stocks.

Climate negotiations. The 2011 Durban Climate Summit resulted in an extension of the Kyoto Protocol and its mechanisms, and a commitment to develop a post-2020 global agreement "with legal force" on climate change policies by 2015. However, if the financial crisis continues, countries and investors may be unwilling to agree on new binding targets or other obligations by 2015, making uncertain the future of international climate agreements and thus the potential impacts on renewables support policies and investment.

Fukushima accident. The March 2011 nuclear accident in Fukushima, Japan, has caused some countries to rethink, slow, or begin to phase out their nuclear development programs. In some countries this could increase reliance on energy efficiency measures and renewable energy deployment. For example, the European Union recently released a Roadmap 2050 that included a low-nuclear scenario, which calls for future shares of renewable energy to increase faster than they would with a larger nuclear contribution. In other cases, however, scaling back plans for nuclear power could actually increase reliance on fossil fuels (with or without carbon storage).

The Arab Spring or Arab Awakening. This wave of demonstrations and uprising across northern Africa and the Middle East helped drive up the price of oil, heightening public awareness and concern about the volatility of fossil fuel prices, while also increasing energy security concerns in many countries associated with heavy reliance on imported oil and influencing policy decisions.

Development of unconventional fossil fuels. As global energy demand and oil prices rise, interest and investment in new fossil fuel resources is increasing: oil from tar sands in Canada, shale gas in the United States, deep-ocean methane hydrates in Japan, and coal-bed methane in many countries. Although many of these resources are not yet applied on a large scale, their development and promise are already having an impact on energy prices, government policies, and investment decisions.

3.5. ADDRESSING ADVERSE AND UNSUSTAINABLE IMPACTS

On balance, modern renewables pose fewer costs to the environment and society than do current energy systems; however, there are costs associated with energy production in all forms. As deployment of renewables scales up rapidly and significantly, the potential for adverse impacts will increase as well—for instance, on land use, soil, air, and water quality, and other factors related to sustainability. Such impacts might, in turn, affect social welfare and public acceptance. It will be important for policy makers at all levels to work to minimize potential adverse impacts.

In general terms, sustainable development is defined in the Brundtland report (1987) as "development that meets the needs of the present without compromising the ability of future generations to meet their own needs". In this chapter, sustainability—or sustainable development—integrates environmental, social, and economic dimensions of energy production and use. These include environmental factors such as impacts on air, soil, and water quality, land use and biodiversity impacts, as well as influence on global climate change; social issues such as related health impacts, land rights and ownership, and labor conditions; and economic issues such as energy security (both availability and distribution of energy resources, and variability and reliability of supply) and the potential for improved energy access.

The most significant and broadest concerns to date have arisen with regard to the production and use of energy from biomass resources. Concerns related to other renewable technologies revolve primarily around the extraction and consumption of raw materials.

3.6. BIOMASS ENERGY AND SUSTAINABILITY

Traditional biomass, used mainly for cooking, poses significant health and environmental costs; this section, however, focuses on modern bioenergy. Depending on choices made regarding biomass feedstocks, production (including land use), and conversion pathways, there exists the potential for negative impacts on soil and water (including contamination, soil erosion, lowering of water tables); air quality; and global carbon and nitrogen cycles. Some studies have determined that, under certain circumstances, net GHG emissions associated with biomass energy (particularly biofuels) might even exceed those associated with fossil fuels. Further, feedstock and cropping choices can also lead to competition with food and fiber needs for humans and livestock. Combustion of municipal solid waste can pose environmental or health risks, and the use of agricultural and forestry residues could negatively affect soil quality or biodiversity.

Other sustainability issues related to biomass energy include socioeconomic concerns such as land rights and labor rights; these issues are not addressed here.

Toward 2050

In a joint 2010 study (*Better Use of Biomass for Energy*, 2010), the IEA implementing agreements for RETD and for bioenergy identified opportunities for improved GHG emission outcomes and policies to promote better bioenergy development. While the potential global contribution of biomass energy that could be produced without negative impacts on biodiversity, soils, and water resources will depend on future developments in the agricultural and forestry sectors, the study found that the technical potential for 2050 ranges from 250 to 500 EJ. This represents 50–100% of current global energy use, or an estimated 25–33% of projected global energy supply by 2050, according to this study (see Figure 3.1).

To achieve this potential in a sustainable manner, policy makers need to address several challenges in areas related primarily to feedstock production—including land use and fossil fuel inputs—and also to conversion routes.

Land use and greenhouse gas emissions

The use of land for biomass production can have significant consequences for life cycle GHG emissions associated with biomass energy. Impacts vary according to feedstocks; for example, whether bioenergy is produced from energy crops (and what types of crops and how grown), agricultural and

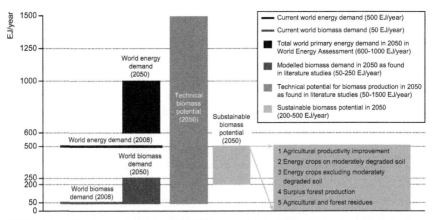

FIGURE 3.1 This figure from the IEA-RETD *Better Use of Biomass for Energy* study illustrates that the upper limit for the available potential of sustainable biomass by 2050 (green bar) is about double the upper limit for the contribution of biomass to a sustainable energy supply (yellow bar). For interpretation of the references to color in this figure legend, the reader is referred to the online version of this book.

forest residues, or from other waste. They also depend on how and where feedstocks are grown. For instance, if existing vegetation (particularly if sequestering high levels of carbon) is cleared to establish energy plantations, the carbon emissions that result could eliminate any potential climate benefit associated with replacing fossil fuels for several years or even decades into the future. In contrast, cultivating perennial (rather than annual) crops on degraded land can increase the soil carbon content, resulting in potentially significant GHG reductions.

Twin pillars of more- sustainable biomass energy

There are two main ways to improve the sustainability of biomass energy: minimize impacts associated with feedstock production and improve efficiency of conversion to useful energy.

In the short term, strategies to minimize impacts on land use might include:

- Use biomass residues and wastes.
- Favor crops with relatively low land demand and a high yield. Be aware of other potentially unsustainable practices, such as the use of genetically modified crops or crops requiring significant fertilizer input.
- Select land that has became redundant due to yield increases.
- Select abandoned or degraded land that is not in competition for carbon sequestration, or for food, feed, or fiber production.

For the longer term, international mechanisms such as the Reducing Emissions from Deforestation and Forest Degradation (REDD) mechanism are being developed under the umbrella of UN climate change negotiations. REDD rewards land owners for increasing sustainability of their land use.

Strategies also exist and are evolving for improving the efficiency of feedstock conversion to useful energy. As conversion processes become more efficient, less feedstock will be required to generate the same amount of useful energy. Converted biomass products from biorefineries could replace fossil carbon, for example, in chemicals, fibers, pharmaceuticals, and plastics. Ideally, the energy content of bio-based products can be recovered (or "cascaded" down) to be used for generating electricity, heat, or transport fuels.

Biomass sustainability indicators and certification

The direct land use and other effects of bioenergy production can be controlled, at least in part, through certification systems that create voluntary or mandatory standards related to life cycle emissions, impacts on biodiversity, and other criteria. Certification procedures are being implemented in the EU and the United States.

Further, government policies can recommend or mandate basic guidelines for better use of bio-energy, as set out in the recent RETD study *Better Use of Biomass for Energy*, 2010.

To improve efficient use of bioenergy:

- Increase the amount of fossil fuels replaced with biomass, measured in terms of energy output per ton of biomass in the case of waste or residues, and energy output per hectare in the case of biomass cultivation.
- Increase the efficiency of traditional cookstoves and heating devices in non-Organization for Economic Cooperation and Development (non-OECD) countries and increase the use and efficiency of combined heat and power plants in OECD countries.
- Encourage investments in energy-efficiency improvements (in energy production, transformation, and end use).

To maximize GHG emissions reductions:

- Set requirements for GHG reductions that apply to the entire bioenergy life cycle and that establish a minimum threshold for emissions reductions—relative to fossil fuels. For example, land-use change emissions should be measured in terms of CO_2-equivalent reduced per ton of biomass or per hectare.
- Provide incentives for bioenergy routes that maximize GHG emissions reductions.

To avoid competition with food, feed, and fiber:

- Promote the cultivation of biomass feedstock on agricultural land set free from significantly increasing agricultural yields.
- Promote cascading use of residues and wastes from biomaterials for energy.
- Develop bioenergy strategies together with a strategy for food security.

3.7. SUSTAINABILITY ISSUES RELATED TO OTHER RENEWABLE TECHNOLOGIES

Most other renewable energy sources do not emit GHGs or other pollutants while producing energy. Nevertheless, other sustainability impacts are associated with the manufacture of technologies and their transportation and installation (e.g., materials use, energy input), drilling to tap high-temperature geothermal energy, changing ecosystems in the case of hydropower, and land and water use (e.g., for photovoltaic cleaning, thermal cooling).

Massive deployment and production of renewable energy technologies will face a number of issues related to materials use, production, decommissioning, recycling, and other environmental concerns, such as impacts on landscape or aesthetics, wildlife, concerns about potential noise and effects on property values, and so on. This section focuses on materials related to impacts. Again, despite these concerns, on balance the impacts of renewable technologies are considerably lower than those associated with fossil fuels and nuclear power.

> Regarding rare earth metals, it will be critical to focus on material efficiency, recycling, and alternatives.

Materials

The manufacture of renewable energy (and associated) devices and appliances requires the extraction and use of raw materials. For example, wind turbines require concrete foundations and steel. Demand for raw materials associated with renewable energy technologies is still modest, but it is increasing. The American Wind Energy Association (AWEA), for instance, projects that steel required for wind turbines in the United States will grow fivefold over the next two decades, from 460,000 tons in 2010 to 2.3 million tons by 2030. Nonetheless, recycling of steel and development of new materials will minimize impacts on national steel production and prices.

Closely linked to some renewable energy technologies is the growing demand for rare earth metals like indium, gallium, cobalt, and lithium. The U.S. Department of Energy estimates that clean technologies like solar cells, wind turbines, electric vehicles, and fluorescent lighting are currently responsible for 20% of the world's total demand for rare earth metals. The mismatch between the limited number of supply countries and growing global demand for these metals has already led to supply disruptions and price increases.

Ecofys has quantified the expected demand for rare earth metals associated with wind and solar technologies following the IEA's Blue Map scenario (see Chapter 2). Growth in consumption of gallium (+55% between 2010 and 2020, without additional "learning") and germanium (+30%) is expected to be substantial, particularly after 2020. It will be critical to focus on material efficiency, recycling, and alternatives if very high levels of renewable growth are to be achieved.

A combination of continued technological advances and government policies will be required to address such challenges.

Geopolitical Benefits and Considerations

Regarding geopolitical issues, international trade, import and export, and sovereignty over resources, renewable energy will have a far lower impact than fossil sources do currently or ever will have in the future. Renewable sources are abundant and every country is blessed with a diversity of renewable energy sources.

As noted by the late Hermann Scheer (former German politician and author), no one can own or have a monopoly over the sun's resources, so there will be no "wars over solar energy." Decentralized harvesting of local renewable energy can benefit communities directly, giving them more control over their energy sources while also creating jobs and contributing to local economic development. However, if renewable energy is harvested in a centralized and extremely large-scale manner, and/or without regard for relevant stakeholders, it may be necessary to rethink this assumption. To ensure security of supply, agreements and transmission lines (particularly those that cross international borders) will need to be "politics-proof"; likewise, it will be important to ensure domestic access to renewable resources and resulting energy services. Similarly, large-scale energy generation at sea could require grids and pipelines that are situated in international waters, calling for international agreements and cooperation.

3.8. TO GET READY FOR THE NEXT STEP

Policy makers have the power to pave the way for rapid and sustainable deployment of renewable energy. While breaking down the barriers discussed in this chapter, policy makers need to consider carefully the developments and trends in the international landscape as they design and implement policies.

Part Two of this book provides examples of policies and policy frameworks that can help to overcome these barriers and analysis of what has worked to date and why.

Policy Experiences and Lessons Learned

Policies to Get on Track: An Overview

Contents

Can we get on track in the next few years to achieving the transition to a clean energy system—one that is based greatly on renewable energy? And, if so, how do we get there?

Part Two builds on the trends, outlooks, and barriers described in Part One to examine the policies and conditions that have driven renewable energy development and deployment in recent years. What policies have been most successful to date and why? And what lessons can policy makers apply in the near future to significantly increase deployment of renewable energy technologies in the next 5–10 years?

Renewable energy technologies offer significant potential economic, security, environmental, and social benefits, as discussed in Part One. The "first movers" among countries or even cities will likely benefit more than those who wait to follow this path and will become the leaders in manufacturing of renewable technologies, creating new jobs and further improvements. However, this calls for mechanisms to remove the barriers and open up opportunities (see Chapter 3). These barriers can affect real or perceived risks and access to financing, the (perceived) cost of capital (investment costs), access to siting and permitting, and other needs for development and access to markets. Policies are required to address and overcome these barriers and can help to create or expand markets.

4.1. DIFFERENT POLICIES FOR DIFFERENTIATED MARKETS AND TECHNOLOGIES

Challenges and barriers differ from one renewable energy resource and technology to the next and from one end-use sector to the next. For example, ocean-based technologies such as wave and tidal energy are less mature than many other renewable technologies and often face challenges related to sea-bed rights, permitting, and grid connection. The wind and sun are variable resources and not always available for dispatch when needed. Some renewable projects, such as geothermal and biomass power, take longer to develop than others and thus require longer lead times. The heating/cooling sector faces a fragmented market with millions of large- and small-scale heat producers and consumers.

Challenges and barriers can also differ according to the stage of renewable energy development in a country or region. While some countries are just starting to adopt or expand renewables focusing on resource studies and public awareness challenges, others are working to develop infrastructure and forge international agreements to integrate large shares of renewable energy into their energy systems. A meaningful and broad acceleration of renewable energy deployment in the next five years will require policies that effectively and efficiently address the range of barriers and challenges that are specific to each situation.

Renewable energy markets, investments, and industries have seen rapid growth in recent years, and some countries have already begun their own energy transformations. Government policies, which are increasing rapidly in number and variety, have played a crucial role in driving these trends and in accelerating renewable energy deployment by helping to overcome numerous challenges and barriers.

To get on the pathways described by the scenarios in Chapter 2, such trends must continue and even accelerate over the upcoming years. Additional and more comprehensive policy packages will be required to level the playing field for renewables by creating markets and driving technology advancement in order to sustain this growth for the long term. Policies that put a price on carbon—whether through taxes or trade—will be important for internalizing costs associated with energy production and use, and for helping to level the playing field for renewables. This will also help to cure the misconception of costs of renewable energy generation, which is still a significant barrier.

Also important in this regard is to consider that government interventions exist throughout the energy sector and have a strong influence on the balance among resources. For example, fossil energy markets around the world still profit from many billions of U.S. dollars of subsidies—estimated by the International Energy Agency at about half a trillion U.S. dollars in 2010. These can be direct subsidies, or indirect incentives such as taxation reductions or exemptions. In addition, the "external costs" of energy production and use, such as health and environmental damages, are generally not included in energy prices. As such, the energy sector is not a free market, and policies for renewable energy should be viewed within this context.

Importantly, policies that focus specifically on renewable energy and that create an enabling environment for renewables are also required, particularly in the near term as countries navigate through the economic recession and work to attract significant resources to renewable energy during one of the worst international credit crises in decades.

4.2. TYPES OF POLICIES—DIRECT SUPPORT

Direct policies for renewables can create markets (demand pull) for renewable capacity or generation through public procurement and policies that spur private demand and investment and through command and control regulations. Policies are important for creating favorable regulatory environments that overcome or eliminate existing barriers to market entry. Through research and development (R&D) policies (supply push), governments can help to create new or improved technologies directly by funding R&D.

Deployment Policies

Policies to support renewable energy directly can include fiscal incentives, public finance, and regulations.

Fiscal incentives include grants, rebates, and tax-based incentives, and they involve either payments from the public treasury or reductions in required contributions to the treasury. Tax-based systems can include tax reductions or exemptions (e.g., from sales, property, or value-added taxes) and help to reduce the total cost of investment. Tax credits, either production or investment, provide an annual income tax credit, meaning that investments can be fully or partially deducted from tax obligations. Indirect fiscal incentives are also possible via exemptions from eco-taxes, carbon taxes, or other levies on fossil fuel based energy.

Public finance policies, including low interest loans, guarantees, and public procurement, provide financial support for which a return is expected, or provide investment security by assuming some financial liability, such as through public–private partnerships. The government shares risk where the private sector cannot or will not act on its own. Fiscal incentives and public finance policies can be applied to technologies at any stage of development, whether in the basic R&D stage or in early or late stages of commercial development.

Regulations are used to target technologies once they reach the marketplace. They include quantity- or price-driven policies (e.g., quotas or feed-in tariffs) for electricity, heat obligations, and biofuels blending mandates. Regulations are rules that guide or control conduct, and they govern factors such as access to markets and networks (through access rules for electricity, third-party access for heating, and blending mandates for biofuels), prices or quantities to be supplied, and the quality of energy to be supplied (e.g., green labeling or green energy purchasing). Increasingly, governments are relying on regulatory policies to advance the deployment of renewable energy, in part because they generally operate independently from the public budget.

Technology Development Policies

Government R&D policies can play an important role in advancing renewable energy technologies at all levels of development, from basic research, across the so-called "Valley of Death" (when technologies are not yet proven so risks are high, but significant increases in investment are required), to commercialization and beyond. The private sector tends to under invest in renewable technologies (as well as many other technologies), because the broader benefits to society cannot be fully captured by the innovator. R&D policy options include public finance policies and fiscal incentives such as public loans or venture capital, academic R&D funding, grants, prizes, tax credits, and public–private partnerships.

R&D policies are most effective when combined with deployment (demand-pull) policies. Used together, they can create a positive feedback cycle by inducing private sector investment in R&D, which can improve performance and reduce cost. This, in turn, expands markets and drives down costs through economies of scale, which attracts further investment and continues the cycle.

4.3. TYPES OF POLICIES—INDIRECT SUPPORT

The surrounding context, or environment, affects how renewable energy technologies will develop and be deployed. It is made up of elements such as (and the interactions of) institutions, infrastructures, broad political agreements and strategies, and various actors (government, finance and business communities, civil society) and circumstances (economic context, land-use and public-perception issues, financing availability, etc.) that can affect motivations, levels of risk, changes in behavior, or other forces that have an impact on the deployment of renewable energy.

Understanding this environment is important for two reasons. First, it helps in determining how best to design renewable energy policies to work within a given context. Second, it is useful to understand because government policies can change this environment, or make it more enabling for renewable energy.

Important elements of an enabling environment (i.e., that go beyond renewable energy-specific policies mentioned above), include:

- Planning and permitting at the local level
- Provision of necessary networks, infrastructures and markets
- Technology transfer and capacity building
- Participation of a broad range of stakeholders, including information sharing, training, and learning from experiences of others
- Integration of renewables policies with policies in a broad range of sectors (including the broader energy sector, as well as agriculture, forestry, education, urban planning, etc.) at all levels of governance

Further, programs that have been long-term and multifaceted (addressing multiple audiences with a variety of support instruments) have been shown to be the most successful. Policies are more likely to be successful if the design of the policy instrument gives appropriate consideration to the maturity of the technology, the existing markets, supply chains (for manufacturing, integration, infrastructure, maintenance), and infrastructure. The markets, supply chains, and infrastructure may vary widely by region; some regions support strong, well-developed markets, whereas others may be in nascent stages of development and therefore require support that is targeted more toward establishment of the necessary infrastructures and supply chains.

4.4. EFFICIENT AND EFFECTIVE POLICIES

Thus far, most government policies have addressed the electricity sector, but the focus of renewable energy support policies is broadening to include the transportation and heating/cooling sectors as well. The next three sections provide brief overviews of deployment policies used in each of these end-use sectors and some lessons learned. R&D and enabling environment policies are not specifically discussed, but they are mentioned in case studies throughout the following chapters.

As the following chapters discuss, some policies—or combinations of policies—have proven more effective and efficient than others in producing a substantial and rapid increase in production of energy from renewable sources to date. To the extent that efficiency and effectiveness are used to analyze the impact of various policies, it is helpful to provide definitions of these two terms:

- Efficiency of a policy is measured by its cost-effectiveness in the short-term, and its ability to drive technology development in the longer term (or dynamic efficiency).
- Effectiveness of a policy is determined by the amount of capacity added/ new generation, or by the extent to which intended objectives are met.

There is some evidence that the effectiveness of policy support is more important during the early stage of market commercialization, whereas market compatibility is increasingly important for more advanced technologies and markets. Whether or not this is the case, there is much evidence that government policies will maximize potential effectiveness and efficiency—to expand and broaden markets, drive innovation, and create domestic industries and new jobs—if they are transparent, clear, and sustained, and if they minimize risk and maximize the breadth and depth of potential benefits of renewable energy.

4.5. ROADMAP FOR PART TWO

Chapters 5 through 7 examine policies used in each of the three end-use sectors: electricity, transportation, and heating/cooling, respectively. Each chapter describes policies used in the specific sector, analyzes their effectiveness, and provides lessons learned from experiences to date in several countries around the world. Addressing end-use sectors separately does not imply they cannot all be advanced simultaneously, something that many countries are already doing.

Chapter 5 includes a detailed discussion regarding fiscal incentives and public finance; while most experiences with these policies have occurred in the electricity sector, the lessons learned can apply across the board. Also included in Chapter 5 is a model policy framework for development of offshore renewable energy technologies. Chapter 6 includes a policy framework for the co-evolution of electric vehicles and renewable electricity.

Chapter 8 reviews the policies and issues related to transformation of entire energy systems. The successful transition to a sustainable energy future will require both the integration of a broader range of renewable technologies into the energy system and a significant increase in the overall share of renewables in energy supply. This chapter explores the greater level of systems thinking that will be required to achieve these aims, including the need for a more holistic view of policy making and a strong linkage of renewables with energy efficiency improvements.

Cities and other local communities have the potential to create change, not only locally but also on the national or even global scale. Chapter 9 focuses on policies and experiences at the local level, looking across all sectors. Lessons learned are relevant at the local level and can be useful for decision makers at higher levels of government as well.

Successful policy frameworks succeed in great part because they help to overcome barriers and challenges to renewable energy technology development and deployment, thereby reducing related risks and drawing investment to research and development, manufacture, project development, infrastructure, and so forth. Over the coming decades, trillions of U.S. dollars in new investment money will be needed to finance massive deployment of renewable technologies on the scale required to achieve the energy transition. Chapter 10 reviews the existing barriers to financing, proposes possible new sources of funds, and examines policies that can be used to improve the risk-to-reward ratio in order to attract these new investors to renewable energy.

Throughout these chapters, a number of detailed case studies look in-depth at local, national, and regional policy experiences and the major factors behind their successes or failures. Note that lessons learned in many of the case studies are relevant to more than one end-use sector, so there may be several links to the same case study across several of the following chapters.

Policies for Power Markets

Contents

Historically, the vast majority of policies to support renewable energy have been focused on the electricity sector, although interest is increasing in policies to advance renewables in the transportation and heating/cooling sectors. A wide range of renewable resources and technologies are available for generating electricity, and government policies need to account for the large variety that they represent with regard to stages of maturity, scales of deployment and generation, potential uses, and circumstances. Most policies directed toward this sector have focused on grid-connected generation, from very small photovoltaic (PV) systems to large-scale wind or hydropower projects.

Some renewable resources are better suited to large-scale centralized application, often requiring transmission of power over great distances (e.g., geothermal and concentrating solar thermal power), while others can be developed in either large, centralized, or small distributed applications and often right at the point of demand (e.g., wind, hydropower, solar PV). Further, some renewable resources can provide dispatchable power, while others provide more variable generation, although some of these (e.g., solar) can be available for valuable peak power production. Distributed systems can be connected to a mini-grid or a larger transmission and distribution system, or can be stand-alone systems producing power in more remote locations.

Beyond applications, there are disparities in levels of maturity and costs. For example, some renewable power technologies are broadly competitive with current market prices, while many others are more expensive but can provide competitive energy services under certain circumstances (depending on available resources, local infrastructure, etc.). Policy frameworks need to be designed with this variety in mind, as discussed in this chapter.

Because the electricity sector has received the most attention thus far, most of the experiences and lessons learned have been in the area of renewable electricity policies. At the same time, electricity is an energy carrier and, as such, can provide a wide range of energy services including mobility (e.g., via electric trains, subways, cars) and heating and cooling. Therefore, the policies and lessons discussed in this chapter can apply directly to the transportation and heating/cooling sectors as well, to the extent that electricity is used to provide relevant services.

This chapter draws on the vast experience with renewable electricity policies, providing a brief discussion of policies used to date, followed by an extensive section with analysis and lessons learned. Policy instruments are structured into the following categories: fiscal incentives, public finance, and regulations. Much of the discussion on fiscal incentives applies to transportation and heating/cooling technologies as well, but it is located here because it is based primarily on experience in the electricity sector. This chapter concludes with a brief section on key recommendations for advancing the use of renewables in the electricity sector. Also included in this chapter is Box 5.1, Model Policy Framework for Offshore Renewable Energy Development.

5.1. OVERVIEW—POLICIES FOR RENEWABLE ELECTRICITY

Fiscal Incentives

Governments generally adopt a combination of policies to advance the production and use of renewable electricity, although some rely primarily if not entirely on fiscal incentives. A broad range of fiscal incentives has been used in the electricity sector, including tax credits, tax exemptions or reductions, energy production payments, grants, and rebates.

In general, fiscal incentives can compensate for the various market failures that act as barriers to renewable energy, and thereby reduce the costs and risks of investing in technologies and projects. They can be used to help steer investment toward specific technologies, uses, or locations.

Tax Incentives

Tax incentives can influence the supply or demand side and are quite flexible (at least in theory) with regard to technologies or markets targeted, and the ability to gradually increase or decrease them as needed. Tax incentives include tax credits (based on up-front investment costs or electricity production), accelerated depreciation, property tax exemptions, and others.

The United States has relied primarily on a federal production tax credit (PTC) (see Case Study 1) and an investment tax credit (ITC) at the national level (also implemented at the state and local level in some cases). Companies that generate wind, solar, geothermal, and "closed-loop" bioenergy (using dedicated energy crops) are eligible for the PTC, which provides a 2.1 cent/kWh benefit for the first 10 years of a renewable energy facility's operation. Other technologies receive 1.0 cent/kWh. Businesses and individuals who buy solar energy systems are able to receive an ITC of 30% through 2016.

Another attractive incentive for investors is *accelerated depreciation*, a tax provision that allows companies to write off some portion, if not all, of the purchase value of qualifying equipment against their profit in the first year(s) after purchase. The accelerated write-off generates important cash flow for companies developing/installing renewable energy technologies, thereby attracting investment money. Examples include the Irish Accelerated Capital Cost Allowance, and a Canadian accelerated capital cost deduction under the Income Tax Act for investments in equipment using a renewable energy source.

Property tax exemptions allow property owners to exempt from their property/land taxes the value that a qualified renewable-energy installation adds to the overall value of the property. Such incentives can be implemented at the local or state level.

CASE STUDY 1 U.S. Federal and State Policies to Support Wind Power

Despite having significant renewable resources, the United States has experienced cycles of boom and bust in the wind industry and has seen relatively little development of other renewable resources. Failure of Congress to consistently renew the federal PTC, which frequently has been reinstated retroactively after its expiration, led to significant dips in new installations in 2000, 2002, and 2004, succeeded by peaks in the years following its reenactment. In contrast, when the PTC has been reauthorized consistently before expiring, the rate of annual installations has experienced steady growth; total capacity increased from just over 9 GW at the end of 2005 to more than 40 GW at the end of 2010. In response, from 2005 to 2009 U.S. domestic manufacturing of wind turbines and their components rose 12-fold.

However, as a result of the economic crisis that began in 2008, reduced corporate tax liabilities meant less tax-related investment was available for renewable energy projects. As a result, the PTC was no longer very effective in supporting renewables development. In response, the federal government stepped up its support for renewables in late 2008, extending the PTC through 2012 and providing project developers with the option of taking cash grants rather than federal tax credits. This policy contributed to a record level of new wind capacity (more than 10 GW) installed in 2009. At the same time, a delay in extension of the PTC until late in 2008 accelerated the drive to complete projects in that calendar year, but slowed the pace of investment in new projects while investors waited to see if the credit would expire at year's end. Without this uncertainty, the 2009 market might have been even larger.

In 2010, the market dropped off again, to about half the 2009 peak, due in great part to lower natural gas prices, which reduced wholesale power market prices, and because a slowing in national electricity consumption reduced demand for new renewable capacity. Installations were up in 2011, and growth in 2012 is expected to continue, but the pending expiration of the PTC at the end of 2012 already casts its shadow on expectations for the next few years. The American Wind Energy Association expects that annual installations will drop to 2000 MW by 2013, instead of a continued growth by 8000–12,000 MW annually in the period through 2016 in case of a four-year extension. As of publication, the U.S. Congress still has to decide on PTC extension (Figure 5.1).

While federal tax credits have helped to improve the cost-effectiveness of renewables, particularly wind power, state level policies have helped to drive growth by providing greater market certainty, thereby influencing the location of development. The most commonly used policies at the state level include tax incentives, interconnection standards, net metering, and quota obligations—known generally as Renewable Portfolio Standards (RPS). Between 2005 and

CASE STUDY 1 U.S. Federal and State Policies to Support Wind Power—cont'd

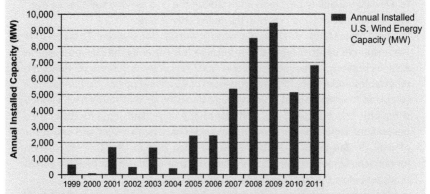

FIGURE 5.1 U.S. Annual and Cumulative Installed Wind Power Capacity, 1999–2009. (*From IPCC SRREN, AWEA and GWEC*). For color version of this figure, the reader is referred to the online version of this book.

2010, the number of U.S. states with RPS policies increased 50%, from 20 to 30 (plus Washington, D.C.). Many states require electricity providers to obtain 20% or more of their power from renewables by 2020; in early 2011, California increased its RPS target to 33% by 2020. As of 2008, state RPS policies required a collective renewables capacity of more than 65 GW by 2020.

The design of RPS policies varies widely from state to state, as have the results. Some states have failed to achieve targets set under their quota/RPS schemes, due to a number of factors including inadequate penalties for noncompliance. In many cases, the sociopolitical context and siting barriers have impeded renewable energy development, demonstrating the need to address barriers such as siting and transmission in addition to establishing targets and financial incentives.

However, other states have experienced rapid growth with RPS policies. Texas first enacted an RPS in 1999 and has consistently surpassed its legislated targets. The state achieved its 2015 RPS target (5 GW installed renewable capacity) six years early and, by some estimates, reached its 2020 target (10 GW) in 2010, when the state of Texas was home to more than one-quarter of U.S. wind capacity. In addition to strong wind resources and federal tax incentives, the key to Texas' success has been state-level support policies such as the RPS, ease of siting and permitting, and standard interconnection procedures. Particularly important is the concept of designated Competitive Renewable Energy Zones with expedited transmission construction, which was developed to address transmission limitations that had slowed the pace of development. In addition, the major utility in the state, ERCOT, has enacted demand-side measures

Continued

CASE STUDY 1 U.S. Federal and State Policies to Support Wind Power—cont'd

to alleviate variability problems and stabilize the grid as wind supplies an ever-increasing share of wind power in the state's energy mix.

As in other states with RPS laws, wind power has dominated in Texas (despite vast resources of other renewables) because it is currently the cheapest renewable technology. Between 1998 and 2007, capacity additions of nonhydro renewables under state RPS laws consisted of 93% wind power, 4% biomass, 2% solar, and 1% geothermal. To encourage diversity, several states have created subquotas or set-asides, particularly for solar energy (and some for distributed generation). Experiences in meeting set-asides have been mixed, and only three of nine U.S. states with such obligations met their 2008 targets. One explanation for nonachievement is that caps on the costs that utilities must bear are sometimes set below the level required to meet the state targets.

In sum, the combination of federal and state financial incentives, state-level policies to establish binding, long-term mandates for renewable electricity, combined with policies to enable access to financing, land (siting), and markets (transmission, etc.), have reduced risk and created greater market certainty, leading to significant investments in wind power deployment and turbine manufacturing. Other renewables have required additional incentives. But steady growth has been possible only when policies at both the state and federal levels have been consistent and stable, providing long-term certainty to investors.

Other tax policies can include reductions or exclusion of renewable electricity investors or generators from energy, pollution, CO_2, sales, and other taxes. Such policies have also been widely used and can directly or indirectly (e.g., CO_2 taxes) benefit renewable electricity by reducing costs relative to electricity from fossil or nuclear sources, making investments in renewable projects more attractive. Examples of countries using such policies include Denmark, Germany, and Sweden (see case studies 8, 12, and 13).

Capital Grants and Rebates

Capital grants and rebates directly reduce the up-front costs of investment in renewable energy systems. Grants are essentially moneys paid up front to help finance an investment, whereas rebates are paid out after the investment has been made (and, in some cases, after specific qualifications have been satisfied).

For example, Japan's Sunshine Program for solar PV rooftop systems provided a combination of net metering (see Section 5.2.2) and capital subsidies/grants that were reduced gradually as prices declined. In the 1980s, Denmark offered a small production payment for each kWh of wind-generated electricity fed into the nation's grid; it was funded partly through a tax on carbon dioxide emissions (see Case Study 13). California and other U.S. states have provided solar PV rebates for installations of eligible systems.

Public Finance

Public finance policies have included concessionary loans, guarantees, and equity investments. Government procurement of renewable electricity technologies and green power purchasing are also options that are gaining significance in some countries.

Public finance policies are important for supporting renewable energy, particularly for technologies in the R&D stage and in developing countries where commercial financing is generally limited. They aim to achieve two objectives: (1) to directly mobilize or leverage commercial investment for renewable energy technologies or projects and (2) to indirectly create commercially sustainable markets for these technologies.

Innovative funding mechanisms are arising at various levels of government, such as the Property Assessed Clean Energy (PACE) loans, which originated in Berkeley, California. Such loans are granted to property owners for energy-efficiency improvements or installation of solar PV systems and are often financed through bonds. They are paid off during the period of use through payments on one's property tax bill, allowing repayment to be matched with energy savings over the course of several years.

5.2. REGULATORY POLICIES

Quantity- and Price-Based Regulations

Increasingly, governments are relying on regulatory policies, particularly quantity-driven (quota obligations and tendering/bidding policies) or price-driven (fixed and premium *feed-in tariffs*; FITs) policies. Regulations that govern access to markets include net metering (and net billing), as well as priority or guaranteed access to the grid and priority dispatch.

By early 2011, electricity *quota obligations* were in force in 10 countries at the national level and at least 50 other jurisdictions at the state, provincial, or regional level around the world, including 30 U.S. states (plus Washington, DC). Quotas, also known as RPS in some countries, generally require that a minimum share or amount of electric capacity, generation, or sales be derived from renewable sources. Policy design varies significantly from one country or state to the next, but in most cases the mandate is placed on a purchasing authority and any additional costs that result are paid by the consumer. Most quotas, but not all, include penalties for actors who do not meet their assigned quota. Increasingly, in U.S. states and elsewhere, quota policies are evolving to include long-term contracts or subquotas for specific renewable energy resources or technologies such as solar PV.

Many quota systems are based on tradable green certificates or renewable energy credits (RECs), as used in Sweden, for example (see also Case Study 8). Obliged parties must obtain a certain amount of these credits to demonstrate that they have met their required quota. The credits and generated electricity are traded separately in the market place, with the price of RECs following supply and demand conditions; as such, RECs represent a potential additional form of income for renewable generators.

Tendering/bidding policies have been used in a number of countries including the UK, China, France, the state of California, the Canadian province of Quebec and, more recently, in Brazil, South Africa, and Uruguay. This process involves a call for tenders to derive a certain amount of generation or capacity from specified renewable sources; bidders compete to win contracts. Tendering allows the market to determine the necessary level of support. All such processes have resulted in project-specific contracts that guarantee prices over a specific period of time, factors that enable a project developer to negotiate favorable financing terms. Under the UK's Non-Fossil Fuel Obligation (NFFO), the government accepted bids from potential generators up to a predetermined maximum level, and successful bids received guaranteed purchase contracts at set prices over a specified period of time. Generators had to install projects within five years or forfeit their contracts.

Whereas quantity-based policies mandate quantity and allow the market to determine price, feed-in tariffs (FITs) set a price for electricity from renewable sources and let the market determine the quantity supplied (with the exception of systems with capacity caps). In addition, they generally guarantee generators with connection and access to the electric grid. An estimated 61 countries and 26 states/provinces had operating

FITs as of early 2011; FITs are also in use at the municipal level in some countries.

Fixed-price tariffs provide specific payments per kilowatt-hour generated (often varying by technology or resource, location, or other factors the government may wish to encourage), while premium payments (e.g., used in Denmark and Thailand) provide either fixed or sliding payments in addition to market prices for electricity that are either linked to electricity prices or set adders; both systems have also been used in parallel (e.g., in Spain). Most FITs include regularly scheduled tariff digressions. They can be funded through tax revenue, through an addition of incremental costs to the rate base, or both, or through other means such as revenues from carbon auctions.

As with quotas and tendering/bidding policies, FITs vary greatly from country to country (and from one state/province/municipality to the next), and thus have different levels of system transparency and complexity, and different impacts on investor certainty and payments, speed of deployment, and ratepayers. For example, under a fixed-price FIT, the combination of network connection, guaranteed purchase, and a set payment create an almost risk-free contract for generators of any size. Premium payment systems expose investors to more risk because generators face some electricity price risk; however, one advantage is that they encourage generators to respond to market price signals.

Access

Access rules govern the ability of renewable (and other) electricity projects to gain access to the electric grid and to the market. It is important for generators to have access to the network and for any renewable electricity produced to be dispatched into the grid; the cost and ease associated with meeting both of these requirements determines the cost and availability of project finance. The ability to connect to the grid depends on rules and regulations at the national, state, and local levels.

As mentioned above, some FITs guarantee both, but not all of them do. Some countries without FITs have guaranteed grid access for renewably generated electricity (the European Union Renewable Energy Directive, 2001/77/EC requires that member states ensure guaranteed connection to the network and injection into the grid).

Regulations that govern interconnection to the grid include technical, metering, contractual, and rate rules that system owners and electric utility companies must follow; such rules play an important role in removing market barriers to renewable energy. Transparent and equitable policies regarding access to the grid and related fees, and simple standardized permitting procedures (especially for small-scale systems), can make investment and development simpler, less time-consuming, and more economical.

Net metering laws are one type of access policy. By early 2011, such laws were in place in most of the United States as well as at least 14 countries at the national level, including several EU countries, Canada, Guatemala, Japan, Jordan, Mexico, and the Philippines. Under this policy, small producers (such as households and small businesses with rooftop solar PV systems) essentially sell excess renewable electricity (beyond their total demand in real time) into the grid at the retail rate. The process generally takes place through a bidirectional meter that allows electricity to flow into the grid when generation exceeds a customer's electricity use, offsetting electricity that the customer consumes at another time. In essence, excess generation offsets electricity that the customer would otherwise purchase at the full retail rate; any net amount exported into the grid over a given time period is generally compensated at a lower rate, if at all. Meters that track *time of use* can reward generators who produce renewable power during peak demand times when electricity is most expensive.

5.3. LESSONS LEARNED—POLICIES FOR RENEWABLE ELECTRICITY

This section reviews the lessons learned from experiences with fiscal incentives, public finance, and regulatory policies in the electricity sector.

Fiscal Incentives

The impacts of fiscal incentives have been difficult to measure because they are generally used as supplementary policies rather than the primary means of support. Some studies have determined that they are most effective when used in combination with other types of policies.

Tax incentives

Tax incentives have the potential to be quite flexible in their ability to target a variety of markets and technologies. To be effective, policies should fit the domestic context; for example, sales tax incentives might not be appropriate where sales tax is low and they should also be made available to nontaxed sectors of the economy. One challenge is in finding a level large enough to increase investment without exceeding the consumer's tax liability and without overdrawing available public funds.

Although they are rarely the only reason for consumers to invest, tax incentives can complement other policies and help to build a sustainable market for renewable energy if well designed, consistently employed, and phased out at an appropriate rate that is long enough to provide stability and jump-start an industry, but without becoming a crutch. The U.S. PTC, for example, has spurred strong growth in wind when in place for a few years running and coupled with state-level polices. Tax exemptions have proven to be important for biomass heat and power in Sweden, for example, and have been most successful where fossil fuel taxes (or carbon taxes) are relatively high (see Case Study 9).

However, countries that have relied heavily on tax incentives have generally struggled with insufficient or unstable markets due greatly to inconsistent (on–off) incentives, particularly when they require annual appropriation of funds. As seen in the United States, growth in the wind industry has stalled when the federal PTC has been allowed to expire or when its extension has been uncertain. The PTC has been most effective when it has been in effect for several years running and in combination with state-level policies (see Case Study 1).

In addition, income tax incentives generally rely on the relatively small number of players with enough tax liability to take advantage of the incentive, thereby limiting potential investors. Experience in the United States demonstrates that tax credit incentives (including the PTC) become largely ineffective during economic downturns, when reduced tax liabilities also reduce the amount of tax-related investment directed to renewable energy projects.

Grants and Rebates

By directly reducing up-front costs of investment, capital grants and rebates and can play a significant role in increasing use of small, distributed projects, particularly emerging renewable technologies.

Neither grants nor rebates require a long-term commitment (as opposed to production incentives), which can be good from a policy standpoint. Grants can stimulate new investors and are often considered most suitable for less mature technologies that face high up-front investment costs. They generally require some form of oversight to ensure that preconditions or specific standards are met, which can increase administrative costs. Rebates work well when their amount is matched with clear targets, tailored for existing policy and market conditions, and used to advance less mature technologies through to mass production. Partnerships with dealers and installers can build consumer confidence, increase competition, and reduce installation costs.

It is important that grants and rebates be designed to complement and work with other policies to address various market barriers (such as zoning restrictions, restrictive interconnection rules, or local building codes). However, as with tax incentives, if funding is inconsistent or uncertain from year to year (as seen with Germany's Market Incentive Program Marktanreizprogramm; MAP, see Case Study 12) it can be difficult for such incentives to provide the stable conditions required for building market awareness, dealer support, and thus consistent market growth.

Investment Versus Production Incentives

Should fiscal incentives support investment or production? Production incentives (e.g., tax incentives or payments tied to production of renewable electricity, heat, or transport fuel) are generally considered preferable to investment incentives because they tend to encourage the desired outcome, which is the efficient, maximum potential generation of renewable energy supply. However, it is important to tailor policies to particular technologies and their stages of maturation.

Generally, investment subsidies (including investment tax credits, grants for investment in renewable systems, accelerated depreciation, etc.) might be most effective for less mature technologies that are still relatively expensive during the demonstration and market introduction stages. By combining other policies with investment incentives, it is possible to avoid unintended outcomes such as low-quality or nonproducing systems. For example, policies paired with technology standards or certification requirements, or provided only for approved equipment or installers, can ensure a minimum quality of systems and installations, while also helping to develop consumer confidence in a technology.

A case in point is India's experience with accelerated depreciation for wind energy. A policy that provided 100% depreciation in the first operating year of new projects led to a wind power boom in the 1990s, but a failure to apply standards or to monitor operating performance and maintenance resulted in relatively low capacity factors. California experienced similar problems in the previous decade thanks to very high tax credits and a lack of standards. In contrast, an early investment tax credit in Germany was accompanied by technical standards for wind turbines and certification requirements that avoided such an outcome.

Public Finance

As with other types of government support, public finance policies should be designed to avoid market distortions that indirectly hinder the development of sustainable, long-term markets.

Loans represent the vast majority of the financing needed for renewable energy projects. Generally, credit lines (from the World Bank and various development banks and agencies) are preferable because they help to build local capacity for financing. Public funds can also be used to buy down the rate of interest, while a commercial finance institution provides most of the financing. Because this reduces the effective interest rate, it lowers the cost of financing and thus project costs. This approach has been used successfully in a number of countries in developed (e.g., Germany) and developing countries (e.g., Tunisia).

Funding mechanisms such as the PACE loans for property owners are considered a positive force when economies are stable or growing. Because they allow borrowers to pay off their loans over a long period of time, rather than up-front when installation costs are incurred, they make financing affordable for even small investors while also overcoming the market failure of split incentives. While programs in the United States have been ordered on hold due to the economic crisis and resulting impacts on the housing sector, the idea has spread to several U.S. cities and counties and was under serious consideration in Germany, Italy, and Portugal as of 2009.

Public procurement is another option with significant potential because many governments account for a large share of a nation's energy use. This can include purchase of green energy (electricity, fuels, etc.) for buildings or vehicle fleets, or deployment of renewable energy technologies. Not only can public procurement help to create a stable market and drive down costs through economies of scale, but it can also increase public awareness and confidence in renewables and set an example for others to follow.

5.4. QUANTITY- AND PRICE-BASED REGULATIONS

There are benefits to both price- and quantity-based mechanisms, and the choice of policy type depends greatly on a government's objectives. There are examples of price-based policies that have been very effective and efficient and those that have not; and examples of quantity-based policies have been effective and efficient, while others have not.

5.5. EXPERIENCES WITH QUANTITY-BASED POLICIES
Quota Systems

Quota systems can result in uncertainties related both to electricity prices and to certificate prices. However, experiences in several U.S. states, British Columbia, and in Sweden show that effectiveness can be high under well-designed quantity-based policies that reduce or eliminate price volatility and reduce risk, and where penalties are enforced and are high enough to ensure compliance with quotas or with contracts under tendering systems. One option is to build a floor price into the market for tradable certificates and to standardize contracts for renewable energy, thereby reducing risk and increasing transparency. Some of the most successful systems have been combined with fiscal incentives, such as the production tax credit in the United States (when in place for several years running) and financial incentives for wind in Sweden.

In general, quota systems have tended to benefit the most mature, lowest cost technologies, and can involve windfall profits for the lower cost technologies; they also tend to benefit incumbent actors. However, quota systems are becoming more effective, as design and implementation improve, based on lessons learned.

Sweden has seen a significant increase in biomass power production under a quota policy. Among European countries with quota schemes, Sweden is the only one that has experienced a high growth rate coupled with relatively low (rather than high) producer profits. A quota policy (RPS) is also greatly responsible for the rapid growth of wind power in Texas over the past decade, although some U.S. states have had less success with RPS laws. Under the UK Renewables Obligation, on average only 65% of each annual obligation was met between 2005 and 2008.

In each of these instances a number of factors played a role, including a lack of adequate penalties for noncompliance, lack of enforcement, application to only a small segment of the market, and caps on the costs that utilities must bear set at too low a level to achieve the target (in the case of U.S. set-asides).

At least one U.S. study has found that RPS quotas have had a significant, positive effect on the development of renewable energy in the states that have enacted them. Additional findings for RPS laws in the United States include requirements that all generation that occurs within a specific state, province, or region may contribute to local economic development, but this could be at the expense of development at the national level. Second, allowing the free trade of RECs can weaken the impact of an RPS.

In addition, subquotas (carve-outs or banding) can be effective in advancing higher cost technologies such as solar, particularly if any caps placed on utility costs are not set too low. At the same time, the use of subquotas can lead to a market separation and narrows the tradable volume within each subquota. In combination with stable fiscal incentives, quotas can address various barriers and provide long-term certainty.

Finally, quota policies are generally considered to have greater market compatibility than FITs, particularly fixed-payment systems, which provide no incentive to react to price signals in the market.

Tendering/Bidding Systems

There has been less experience with tendering/bidding systems to date, and only limited success. The UK's tendering system (the non-fossil fuel obligation, NFFO) described above did not succeed in adding much capacity, in part because limited availability of funds created intense competition to drive down bids, and there were no penalties for noncompliance. In China as well, bidding for wind projects created concerns about price competition and low profitability of renewable energy facilities (e.g., wind farms); at the same time, the program succeeded in bringing a number of large projects on line. China's success has resulted from a variety of policies addressing both the supply and demand side, including a mix of "carrots" (incentives) and "sticks" (mandates) (see Case Study 2).

CASE STUDY 2 China's experience with on- and off-grid renewable power

At the beginning of 2011, China led the world and ranked in the top five for renewable energy production or capacity in all end-use sectors. China accounted for 50% of global wind capacity additions in 2010 (up from 4.4% in 2005), was the leading market for solar thermal systems, and the third largest producer of ethanol. Including hydropower, China had more renewable power capacity than any other country; not including hydro, China was second after the United States. During 2010, China led the world for investments in renewable energy (USD 50 billion). The country plans to spend between USD 400–600 billion in the power sector over the next 10 years.

China first subsidized renewable energy in the 1950s, focusing on small-scale hydropower in rural areas; later on, rural programs expanded to include wind, solar, and biomass energy. Efforts such as the Township Electrification Program have successfully and rapidly brought electricity to millions of rural Chinese due greatly to training and capacity building, education of local and national decision makers, technical and implementation standards, and community access to revolving credit.

Especially since the 1980s, China has pursued a broad-based economic strategy with many integrated elements, with a major emphasis on product export. In the 1980s, China identified clean energy technologies as part of its long-term "State High Tech Development Plan" for new technology innovations. Stable, long-term policies for promoting renewable energy demand are key. Local officials are increasingly being held responsible for meeting official targets and abiding by national regulations. Also, the government supports worker training programs. In some cases, financial support for businesses is provided to send workers abroad (e.g., to Germany and the United States) to acquire technical skills in the wind and solar industries.

Most grid-connected development of nonhydro renewables has occurred since the national Renewable Energy Law took effect in 2006. The law created a national framework to support renewable energy and institutionalize several support policies with the aim of removing market barriers and creating markets for renewables, establishing a financial guarantee system, and creating awareness, skills, and understanding. It has been followed by a number of regulations and measures, including mandates, penalties, and economic incentives to motivate companies to comply with national plans and targets; in addition, the government implemented resource surveys, R&D support for manufacturers, technology standards, grant programs for various renewables, and building codes for the integration of solar water heating systems

CASE STUDY 2 China's experience with on- and off-grid renewable power—cont'd

into new buildings. FITs were established for wind and biomass power plants, while bidding procedures were used for off-shore wind power plants, and later for solar power. In addition, the government mandated capacity quotas for China's largest electricity-generating companies, to be achieved by 2010 and 2020. As China's capacity rises, increasingly aggressive future targets have been established.

China's bidding program has led to concerns about price competition and low profitability of ownership of renewable energy projects, but several large wind power projects have resulted, and some level of price transparency has emerged that has been helpful in setting FITs. Recognizing that this system was a barrier to profitable development, the wind tender was replaced in 2009 with an FIT, with fixed tariffs that vary by regional wind source; FIT tariffs for solar PV were announced in mid-2011.

China continues to face challenges related to interconnection with the electric grid, testing, and certification, due greatly to the rapid pace of development. Approximately 18.9 GW of wind capacity was installed in 2010 but, by the end of the year, about 13 GW had not yet been commercially certified and about 2 GW were not yet feeding electricity into the grid.

The Chinese government continues to learn from experiences at home and abroad, and to address challenges as they arise by revising policies, enhancing technical skills, establishing institutions to support renewable energy R&D, extending transmission capabilities, creating a domestic market to stimulate demand and minimize over-reliance on overseas markets, and establishing a national industry association to bridge the industry and policy-making processes and coordinate development. China has rapidly scaled up its renewable energy deployment and begun to develop a strong domestic manufacturing industry with a mix of policies—supply- and demand-side, mandates, and incentives—in a coherent manner and with a long-term view supported by increasingly aggressive targets.

Tendering schemes can lead to cost-effective support, in theory, because investors must compete for government support. However, tendering poses risks since developers must prepare bids without guarantee that they will be successful. Policy design can reduce but not eliminate these risks; for example, through generation-based tendering (rather than investment based), which involves long-term contracts for electricity produced (see Case Study 3).

CASE STUDY 3 Brazil's Renewable Energy Tender

Hydropower accounts for the vast majority (2010: 79%) of Brazil's electric capacity (totally 110 GW). The national energy development plan (Plano Decenal de Expansão de Energia 2020, published in 2011) foresees an increase of renewable energy projects from about 92.5 to 142 GW by 2020. Although most new capacity will likely be hydropower, significant growth is expected for wind and biomass power as well. The Brazilian government relies heavily on an auctioning or tendering system. Producers and utilities enter long-term power purchase agreements such as in the case of biomass for 15 years, and in the case of wind power for 20 years. Tariffs are determined in the auctioning process.

During the August and December 2011 auctions, power contracts for wind energy were signed at prices below bidding prices for natural gas, showing the increasing competitiveness of wind power in Brazil. The National Electric Energy Agency signed contracts for about 1 GW of wind for a selling price of USD 55/MWh in December, USD 12/MWh lower than the price established in August.

There are two main variants of auctioning. The first counts toward the nation's energy matrix, and distribution companies are the buyers. The distribution companies commit to developing the winning projects and completing their connection to the grid within three years.

The second variant is "reserve" tenders, which are intended to increase Brazil's energy security by adding "reserve" energy to the national interconnection system. The reserve was effectively a way to set aside tendering capacity for large hydroelectric facilities. Companies with winning tenders will sign a 20-year purchase and sale of energy agreement, valid from 2012.

Brazil's tendering mechanisms demonstrate the effectiveness of the Brazilian Government to encourage significant participation in the renewable energy market with private sector finance. However, the long-term success of this tendering process will become apparent in a few years as there is risk, similar to that experienced by the UK with its offshore wind farm tender, that not all winners will develop their awarded projects.

5.6. EXPERIENCES WITH PRICE-BASED POLICIES—FEED-IN TARIFFS (FIT)

Germany has seen strong growth in a variety of renewable electricity technologies in response to its FIT, which many consider to be the model FIT. Germany's FIT, which has evolved over the years as the market has developed and as policy makers learn from experience, has encouraged significant investment in renewable power technologies by minimizing risk. The FIT has not only created broad public support for renewables but has also enabled and encouraged a wide variety of investors—from farmers and homeowners to businesses—to put their money into renewables, thereby expanding the investment pool. The FIT has played a major role in making Germany a renewable powerhouse, and it is credited with helping to drive down the costs of some renewable technologies on a global scale. Denmark, Spain, and several other countries have seen success with FITs as well, and this policy type has spread to a number of countries across Europe and elsewhere, including several developing countries such as Algeria, Indonesia, Nicaragua, Syria, and Thailand (see Case Study 4).

CASE STUDY 4 Thailand and grid-connected generation

In Thailand, decentralized and grid-connected renewable energy, especially biomass, has made a rapidly growing contribution to the country's energy supply. Thailand's renewable energy policies have also helped to promote the development of integrated biorefineries for rice and sugarcane, allowing for simultaneous production of power and heat along with food, fertilizer, and ethanol.

Strong growth in the market is thanks to good renewable resources, including agricultural wastes, and a comprehensive set of policies including streamlined utility interconnection requirements for small generators, tax reductions and exemption of import duties, technical assistance, low-cost financing (including low-interest loans and government equity financing), and an FIT.

Growth in the market has been particularly strong since introduction of a national FIT premium payment in 2006. The FIT provides an adder, on top of the utilities' avoided costs, that is differentiated by generator size and technology type, and is guaranteed for 7–10 years. Under the policy, utilities are provided further incentives to accommodate generators up to 1 MW. Additional subsidies per kWh are provided to projects on mini-grid systems that offset power from diesel generators. To avoid negative impacts on some of the country's poorest people, the government subsidizes the cost of electricity for small consumers.

Thailand offers an example of how FITs can be adapted to meet developing country situations and needs. While the FIT premium has played a significant role in advancing a range of renewable technologies, also significant have been financial incentives for utility companies to purchase power from very small producers. These incentives have encouraged utilities to address the various challenges to interconnection, grid operations, and billing that can accompany distributed generation. By starting with regulations that govern interconnection and adopting FITs only after utilities had gained some experience, the government enabled them to learn by doing as programs and capacity ramped up. Also important, the government has kept an eye on the market and learned from experience, working to address challenges as they arise.

FITs can effectively attract significant investment in new capacity if the FIT tariff is sufficient to meet investors' needs and other major obstacles are addressed. However, FITs have not always succeeded, due in part to design elements as well as additional hurdles to renewables development including high administrative barriers. For example, an early FIT in France led to a wave of applications for wind farms, but relatively few were actually built due to the FIT's capacity limits, turbine spacing rules, and onerous building approval procedures. Italy's early pricing law was not very effective due to a lack of quality standards, difficulty in obtaining financing, problems in accessing the grid, and lack of confidence in continuity of the policy.

Spain's FIT had a rocky start and, although it has spurred significant development in wind power and solar PV, it has recently faced further challenges, particularly for solar PV, demonstrating that even FITs can bring uncertainty and temporarily high per unit costs if policy adjustments are frequent and unpredictable (see Case Study 5). In fact, developments in Spain have increased perceived political risk for all FITs, while significantly affecting the renewable energy industry in the short term.

CASE STUDY 5 Spain's Experience with Solar PV

In 1998, Spain established FIT tariffs for solar PV that were similar to those in place in Germany. Despite having better solar resources, however, Spain installed little new capacity in the years that followed while Germany's market took off. A lack of regulations governing grid connections meant that utilities set their own, often exorbitant rates. National technical standards were established for grid connection in 2001, but a cumbersome registration process for PV producers (including households and small businesses) to sell electricity into the grid, discouraged investment. Regulations to overcome such barriers and minimize risk, along with additional incentives such as investment subsidies and low-interest loans, were needed to launch Spain's solar market.

Spain revised its FIT in 2004 and again in 2007, when payments that were tied to the prevailing electricity price were replaced with guaranteed fixed-price contracts for PV projects up to 50 MW and for 25 years. As a result, newly installed capacity increased nearly fivefold in one year, from 107 MW in 2006 to 555 MW in 2007. By September 2007, Spain had surpassed its 400 MW target for 2010, triggering the countdown to an automatic (but as yet unspecified) revision of the country's solar policy to take place one year later.

Because solar projects do not require much lead time, this expected revision led to a rush of PV development. In addition to a significant increase in truly small-scale PV projects, developers strung together multiple smaller projects to

CASE STUDY 5 Spain's Experience with Solar PV—cont'd

take advantage of higher tariff rates and the economies of scale associated with larger ground-mounted projects. Nearly 2,600 MW of new capacity was installed in 2008 alone, representing about 40% of the global market, and putting tremendous pressure on the Spanish government budget (taxpayer funds were used to cover costs above a fixed retail electricity rate).

The 2008 revisions set an overall 500 MW annual cap (to adjust automatically depending on the previous year's installations), with separate caps for ground-mounted and rooftop projects, and differentiated tariffs (highest for building-integrated PV). The aim was to provide long-term predictability, encourage declining investment costs and greater competitiveness, better control the cost of the FIT, guarantee profits that were more appropriate for a regulated market, and to encourage distributed generation through building-integrated PV. While the result was a significant increase in rooftop projects, uncertainty lasting for months about the design of the new framework, lack of experience with new administrative procedures, and the cap on ground-mounted systems all led to a sharp drop in total installations. A change to a system of tenders to allocate the annual market cap under the FIT, and introduction of new procedures making it more difficult to obtain necessary permits for grid connection, also hindered market development. Spain added an estimated 145 MW in 2009 and 370 MW in 2010, well below the annual cap.

In late 2010, the Spanish government introduced two new laws that represent retroactive reductions in PV support (by introducing a cap on the amount of electricity that owners of already existing installations can sell to utilities). This led to an outcry from the industry and investors claiming that rates of return on existing projects would be dramatically eroded, possibly causing massive bankruptcies, and that future investments in the technology internationally or in Spain could be severely affected. In response, international investors and fund managers with approximately USD 4 billion (€3 billion) worth of Spanish PV projects warned the government of potentially massive losses for their clients, and two EU Commissioners urged the government to ensure that renewable policies remained stable and predictable. Investor confidence was shaken, not only in Spain, but also in other countries where governments were revising renewable incentives in response to the continued economic crisis. As of publication, the Spanish government faces a lawsuit from investors on an international level and from renewables associations on the national level.

In November 2011, the departing government approved a Royal Decree that regulates auto-consumption for homes and small and medium enterprises, paving the way for distributed power generation by simplifying the administrative process for small power-generation facilities up to 10kW. Details were still outstanding as of end of 2011, but the news was welcomed by the renewable energy

Continued

CASE STUDY 5 Spain's Experience with Solar PV—cont'd

sector. However, in January 2012, the Spanish government suspended all support for renewable energy as part of an effort to manage the national debt crisis.

*The dramatic rise and fall of Spain's solar market underscores the impor-
tance of a combination of support policies to overcome regulatory and other
barriers, particularly when a market is immature; the need for stable, predictable
policies to minimize risk; and the importance of careful and flexible policy design
that allows for quick reaction but moderate, transparent, and predictable policy
adjustment when the situation requires it. Spain's experience demonstrates the
importance of spreading the costs of FITs (or other regulatory policies, including
quotas) across the consumer base without relying on the government budget. It
is also important to avoid loopholes—for example, by monitoring applications
closely to prohibit developers from combining projects—and to enforce policy
provisions to ensure that developers do not game the system and increase costs.
Finally, retroactive policy changes must be avoided or governments risk under-
mining years of effort and expense invested to build a strong domestic market,
risk creating widespread negative impacts on renewable energy industries, and
will likely need to go to greater expense at a later date to regain trust and reduce
risk enough to tempt investors back into the market.*

Historically, FITs have been determined to be the most effective and
efficient regulatory option for accelerating deployment of a range of project
sizes and a variety of renewable energy technologies, and have done so at
lower cost per unit of generation. This is due greatly to higher investor certainty provided by FITs, which reduces the cost of capital (required return
on investment) and stimulates investment. The cost efficiency of fixed tariffs
is generally higher than that for premium tariffs due to lower risk premiums
(and thus lower overall costs), but this depends greatly on system design.
The use of well-designed FITs can avoid the need for a host of additional
subsidies.

FITs, particularly those with properly designed and implemented tariff digression mechanisms, are more successful than quotas or tenders at
driving dynamic efficiency, or helping to achieve cost reductions through
creation of a competitive industrial sector. Because more efficient producers obtain greater rents by reducing costs, investment money is assigned to
investors accordingly, therefore, FITs provide dynamic incentives to reduce
the long-run marginal costs of a range of renewable technologies. FITs are
also considered to encourage competition among manufacturers, helping

to drive down costs, and to bring new players into the market. In addition, FITs generally favor local ownership and control of systems, resulting in wider public support for renewable energy and unleashing significant amounts of capital for renewable energy projects.

However, FITs can also have distributional impacts on poorer consumers or on industries with high electricity demand without efforts to address such impacts in policy design. Some have argued, for example, that Australia's residential solar FITs have proved to be a regressive form of taxation, benefiting wealthier households while putting an additional burden on poorer consumers. However, Thailand offers one example of a means for addressing possible equity issues (see Case Study 4). Another argument is that since changes in electricity prices do not affect payment to generators in fixed FIT systems, there is generally no incentive to react to price signals and match supply to demand. In addition, although domestic production requirements can support the local economy and strengthen the political case for FITs, they can involve restraints on renewable energy trade across borders, thereby raising concerns about distortions in trade.

Despite the benefits associated with FITs, there is increasing concern about potential overpayment to generators under FIT policies, and resulting overstimulation of the market. FITs need to provide adequate financial incentives while avoiding generous profits; this requires a careful balancing act, and poses a particular challenge if/when costs are declining rapidly, as seen recently with solar PV. Rapid reductions in PV costs have led several countries to reduce payments for PV or even to stop deployment. Some countries have capped annual payments to minimize system costs, but caps can cause stop and go in the marketplace and unpredictability about qualification for FIT payments, increasing instability for investors. Depending on the cap set, it can also limit the ability to achieve economies of scale in the domestic market.

Another option is a growth corridor (e.g., Germany), in which tariff digressions for new capacity increase automatically if a capacity target is met in a given time period, or decrease automatically if deployment is below the target. Such a system cannot contain costs on its own, but can prevent an overheated market if well designed; at the same time, however, it reduces investor stability because it is difficult to predict when the cap will be reached. Further, adjustments that are based solely on the quantity of capacity installed may not coincide with price trends.

Experience will have to bear out what options work best and under what circumstances.

While the global recession has increased governments' concerns about the costs of FITs, at least one study has found that investors continue to prefer FITs over all other policy options. In 2011, a survey examined the renewable energy policy preferences of 60 U.S. and European clean technology venture capital and private equity investors. Although investor support for most policies had decreased by 2011, particularly in Europe, there was no significant change in the preference of investors for FITs, and FITs remained the most popular policy for minimizing risk, despite recent tariff reductions and uncertainty surrounding the FIT in Spain.

5.7. KEY ELEMENTS FOR SUCCESS

Both price- and quantity-based policy mechanisms continue to evolve as governments draw on experience to adjust policy details and meet their changing needs. Ultimately, policy design and implementation are critical in determining the effectiveness and efficiency of various policies. Effectiveness of quota obligations appears to be improving with advances in design and implementation, although most literature continues to conclude that overall FITs (particularly fixed FITs) are more effective than quantity-based policies due primarily to the fact that they provide greater certainty for investors.

Table 5.1 summarizes many of the key elements of successful quantity- and price-based policies to date.

TABLE 5.1 Key elements of successful quantity- and price-based policies

Quota	Tendering/bidding	FIT
Clearly defined rules of eligibility (resources, actors)	Clearly define rules of eligibility (resources, actors)	Obligation for utility purchase
Adequate penalties (and enforcement) for noncompliance	Adequate penalties (and enforcement) for noncompliance	Priority access and dispatch
Technology-specific subquotas to advance less mature, more expensive technologies	Technology-specific subquotas	Clear standards for grid connection and procedures for allocating transmission and distribution costs

TABLE 5.1 Key elements of successful quantity- and price-based policies—cont'd

Quota	Tendering/bidding	FIT
Long-term contracts	Long-term contracts	Tariffs that are: (1) differentiated by technology and project size, and based on cost of generation (to cover project costs plus an estimated profit); (2) guaranteed for long enough to provide adequate rate of return (importance of carefully calculated starting value)
Consideration of a closed-envelope bidding process	Consideration of a closed-envelope bidding process	Eligibility for all potential generators, including utilities
Minimum payments to ensure adequate return and financing; ensure that prudently incurred compliance costs can be recovered in electricity rates	Minimum payments to ensure adequate return and financing	Regularly scheduled short-term tariff adjustments and longer term design evaluations; consideration of growth corridors rather than caps on capacity
Application to all load-serving entities	Generation-based (rather than investment-based) schemes	Integration of costs into rate base
Clear focus on new capacity; well-balanced supply–demand conditions		Consideration of vulnerable consumers (such as low-income, large electricity users)
Where RECs are used, rules and tracking systems to avoid double-counting of RECs		Tariffs that are stepped by time of day to encourage generation when most needed
No time gaps between one quota and the next		Ease of application and administration
Long-term targets with interim targets that ramp up steadily over time		

Access and Integration

Access policies that are most effective at supporting investment in renewable electricity are those that guarantee both a connection to the network and priority dispatch into the grid (even in the event of a constraint) or that compensate renewable generation when curtailment occurs. Access policies are often linked to regulatory policies discussed above. In Brazil, government regulations have provided sugar mill operators with access to the electric grid so that excess biomass electricity can be sold through contracts or auctions. An increasingly employed regulation for providing access to small-scale producers in particular is net metering.

Net metering policies can play an important role in removing market barriers to renewable energy (particularly distributed systems). Net metering largely means that a (small) generator—for example, the owner of a rooftop PV system—is refunded for the amount of supplied power at approximately the same price as the commodity price, which increases the return on investment for a generator. Net metering can benefit both utility and customer if there are enough systems to affect electricity supply; however, if a large number of such systems are installed, there is the potential that the utility could sustain significant losses in revenue. Net metering policies are most effective when they are available for a broad base of customer classes, do not put a cap on total capacity or on individual system size, and provide credit for net excess.

At least in the United States, studies have concluded that net metering alone does not provide enough incentive to stimulate substantial growth. At the same time, a 2009 study prepared for the National Renewable Energy Laboratory found that U.S. states that implemented net metering policies in 2005 had far more generation from renewable sources in 2007 than those states that did not. Most likely the combination of net metering and other incentives is driving markets in these states.

In the Netherlands, net metering of small customers/generators has been legally mandatory for all utilities for several years, up to a level of 3000 kWh annually. In practice, some utilities allow for net metering until 5000 kWh, which has proven to be a stimulus for small consumers/producers.

Net metering combined with *real-time pricing* of electricity can help to address barriers related to unfavorable power pricing rules. Real-time pricing values renewable electricity (or other distributed generation) at the actual cost of the utilities' avoided generation at any particular time of day. In such cases, customers receive high rates if they sell power into the grid during peak times, which often coincide well with peak generation from solar PV systems, for example. Real-time pricing is not commonly used, but offers the potential to provide significant incentives for some grid-connected distributed renewable generation.

A significant challenge for the *integration of increasing shares* of renewable electricity, in particular, is coping with the variability of some renewable sources such that the system as a whole can match supply with demand. Governments can address this through a variety of avenues:

- Policies can support technological and geographic diversity, which helps to smooth the effects of variability.
- Regulations can ensure that aggregate renewable energy production data are incorporated into operations of the electricity market. Spain, for example, mandates the aggregation of all data from the nation's wind power plants in delegated control centers, which requires online communication with the National Renewable Energy Control Center. Renewable energy projects larger than 10 MW must provide daily supply forecasts to the regional system operator; these can be updated up to one hour before delivery.
- Requirements that the balancing of electricity take place as near to "real time" as possible, such as one hour ahead rather than several hours or one day ahead, can help to accommodate increasing shares of variable renewable sources.
- Demand-side management policies such as real-time pricing can address unfavorable power pricing rules and shift demand to off-peak times.
- Fiscal incentives or public finance policies can be used to increase storage capacity.
- Policies that promote the use of more flexible thermal generation can also help to address variability.

5.8. RECOMMENDATIONS FOR THE ELECTRICITY SECTOR

To date, FITs, if well designed, have been found most effective and efficient at increasing installed capacity of a range of renewable technologies and in creating broad-based public support for renewable energy. However, badly designed FIT systems could experience high costs. Also, there are increasing concerns that FITs may not be politically and economically sustainable in the long term, as capacity increases significantly along with associated costs. However, this is not of concern for most countries over the next five years, and technology costs are declining such that the costs of supporting renewable energy (per unit of capacity or generation) will fall as well. If FITs account for these declining costs, they can drive deployment of renewable energy technologies in power generation quite efficiently.

No matter the specific type of policy instrument implemented in support renewable electricity technologies, there are a number of design elements that may facilitate their effectiveness and efficiency. These include:
- Establishing clear, consistent targets
- Implementing plans and programs to achieve targets
- Monitoring and reporting progress toward these targets
- Designing and implementing evaluations to assess success
- Refining targets and programs based on evaluations or changing conditions

Priority should be placed on well-designed regulatory policies that are transparent and easy to administer while minimizing investment risk and attracting capital. At the same time, it is important to have a combination of policies, including guaranteed access to the grid, standard rules and minimal connection costs for developers (particularly small-scale), and anticipatory transmission planning in preparation for future development.

Creation of development zones is also critical for minimizing risks related to project siting. Such zones should be determined in collaboration with (or by) local planning boards and relevant stakeholders. In general, policies should encourage and enable broad stakeholder participation, including local ownership, particularly to maximize awareness and benefits to local communities and to minimize opposition to renewables.

As seen in Sweden with biomass heat and power, Germany and Denmark with wind and other renewables, public involvement in the planning process and broad public ownership and engagement have maximized benefits and public support. Broad public support for renewable energy will become more important in the future as wind turbines, biomass power plants, solar PV, and other technologies are developed in greater numbers and become more visible and widespread. Policies that encourage utilities and help them to address challenges of integrating renewable energy are useful ingredients for a smooth introduction, as seen in the case of Thailand.

It is important to consider the interplay among the different levels of policies. International agreements can positively influence the implementation of national policies, and the success of national policies may be influenced or aided by regional or state/provincial programs that provide additional support for specific technologies—assuming policies enacted at various levels work in concert and do not contradict each other. Even where no national policies are in place, subnational (e.g., state, province or municipality level) policies can strongly influence the development of markets and supply chains, and in some cases can (and have) pave the way to national level policies.

Finally, policies must be stable, predictable, and long term to minimize risk to investors and ensure long-term growth of renewables. Experience has shown that sudden and unpredictable changes, even to FITs, can result in capital flight away from an industry or country.

See Box 5.1 for a model framework for development of offshore renewable electricity technologies.

BOX 5.1 Model Policy Framework for Offshore Renewable Energy Development

The Renewable Energy Technology Deployment report *Accelerating the Deployment of Offshore Renewable Energy Technologies* (2011) proposes a framework for how to foster deployment of offshore wind as well as wave, tidal, and other ocean technologies. The following is a brief summary of guidelines based mainly on this report.

1. **Market creation.** Strategic and flexible support mechanisms should aim to create a market for offshore renewable energy technologies and should be appropriately sized for the scale of risk and the government's desired capacity level. Policy options include:
   a. Phased tariffs that are higher at early stages of development and decline over time with learning and falling costs.
   b. Flexible tariffs that vary according to distance from shore, water depth, or other factors that affect development costs.
   c. Tender model that calls for bids at specific sites.
2. **One-stop efficient and transparent permitting process.** Permitting rules should be clear and easy to understand, with a single government agency responsible for addressing all permitting issues and requirements. Government should set aside specific zones for development. Licenses for development and operation should require clear milestones for achievement and expiry dates.
3. **Ease of grid connection.** There should be clear rules regarding grid connection (who can be connected, in how much time, at what cost and to whom), and connection should be made available in a timely manner. Adequate recourse should be available for developers if connections are not provided as required.

Continued

BOX 5.1 Model Policy Framework for Offshore Renewable Energy Development—cont'd

4. **Public investment in projects and infrastructure.** Governments should provide fiscal incentives or public financing to assist with early investment, partner with industry to develop demonstration projects, or other measures to signal to lending community that the government is fully behind the offshore renewable energy industry. Support should diminish as the industry matures.

5. **Create an enabling environment.** Development will also be greatly assisted through the following:

   a. **Infrastructure support.** Provide funding support for (or develop) suitable manufacturing bases, foundations, harbors, and other necessary infrastructure;

   b. **Resource and feasibility studies.** Conduct resource studies and basic feasibility studies.

   c. **Knowledge building and sharing.** Create centers of excellence and promote networking and knowledge transfer.

   d. **Capacity building.** Identify shortage of skills and develop programs to attract and train workers over required timescales.

   e. **Guidelines and standards.** Develop clear environmental, technical, health, and safety standards and requirements.

Government demonstration of strong support for the policy framework and long-term commitment to development of the industry should minimize risk and provide the sufficiently attractive returns required to encourage investment and launch the offshore renewable sector.

Transportation Policies

Contents

Three routes exist for increasing the deployment of renewable energy technologies in the transportation sector. Currently, the most important route is the replacement of fossil fuels by renewable transport fuels (liquid and gaseous) produced from biomass. The other two routes involve the use of renewable electricity to charge batteries for electric vehicles, and conversion of renewable energy to hydrogen for use in fuel cell vehicles. Policies to advance renewables in the transport sector have focused on the fuels, vehicle technologies, and related infrastructure, with most attention to date on biofuels.

Biofuels have been produced at commercial scale since the 1970s, and are widely available at fueling stations in a number of countries, including Brazil, Sweden, and the United States. In 2011, liquid biofuels accounted for about 3% of global road transport fuels, according to the REN21 *Global Status Report 2012*. Biofuels are used primarily in the area of road transport, but there is potential in the marine transport sector, and increasing attention to the potential for fueling aviation as well. Limited but increasing quantities of biogas are also being used to fuel trains, buses, and other vehicles, particularly in a number of northern European countries.

Commercialization of advanced, sustainable biofuels, with the help of policies, will be critical for growth and will have to be accompanied by a considerable expansion in production capacity. In the near-term, policies that support sustainable biofuels production from feedstock to fuel (e.g., without affecting food output or creating negative climate or other environmental impacts) will be critical.

Renewable Energy Action on Deployment
http://dx.doi.org/10.1016/B978-0-12-405519-3.00006-2

The use of electric motors running on power derived from renewable sources is still narrow but gaining attractiveness. In addition to the more conventional electric means of public transport, such as trains and subways, interest in private electric transport has increased over the past 2–3 years in parallel with the development of an increasing number and variety of commercial electric vehicles (EVs). In the future, a large number of vehicles may be powered by fuel cells using renewably produced hydrogen; due to their innovative and still marginal role expected for the foreseeable future, hydrogen vehicles are not included in this chapter.

Part of the appeal of personal EVs originates from their zero-emission and low-noise characteristics, which are particularly attractive in densely populated areas. According to the International Energy Agency, the combined international targets announced by mid-2011, if achieved, would result in the sale of about 7 million EVs annually by 2020, with a total of approximately 20 million electric vehicles on the road by then (or about 2% of global vehicle stock). Electrification is not likely to be possible for most marine uses, aviation, or long haul trucks due to driving range and weight limitations or payload requirements.

While biofuels are subject primarily to national or even international policies, early market deployment of EVs is of interest primarily at the regional and city levels due to their local environmental benefits, relatively limited driving range, and more manageable infrastructure requirements (recharging stations, etc.). However, some linkages exist between these routes. For instance, biomass can be used to produce both electricity and biofuels, while some hybrid vehicles can run on both (bio)fuels and power, generated in or outside of the vehicle's engine (see Appendix B). These linkages may enhance the parallel evolution of renewable energy technologies for both transportation and power production.

This chapter covers biofuels-related policies in use with a brief discussion of existing policies related to flex-fuel vehicles (FFV; which can run on any blend of ethanol or gasoline and are necessary for ethanol blends above the 5–15% range) and EVs. Policy instruments are structured into the following categories: fiscal incentives, public finance, and regulations. The following sections first set out the policy options used to date, and then provide analysis and lessons learned, concluding with a brief section with key recommendations for the transportation sector. Box 6.1, Policy Framework for the Coevolution of EVs and Renewable Electricity connects renewable power to the use of electric vehicles.

BOX 6.1 Policy Framework for the Coevolution of EVs and Renewable Electricity

The coevolution of EVs with renewable electricity can advance the transition in the transportation sector while simultaneously helping to integrate variable renewable sources. There is no specific set of policies that will work in all countries and under all circumstances. This is because energy systems and transportation characteristics and requirements differ from place to place depending on the existing political framework, economic situation, existing infrastructure, settlement patterns (e.g., density), modes of transport, and even the climate (which can affect battery life). Thus, it is important to keep in mind regional situations, limitations, and challenges when developing a policy package. However, there are common elements to a potential strategy for the way forward.

RETRANS-2, a report prepared for the Renewable Energy Technology Deployment in 2011, outlines a proposed policy framework for the coevolution of EVs and renewably generated electricity. It is based on a two-phased approach that begins with initial preparation of markets and infrastructure, and moves on to increased deployment of EVs for mass markets:

1. Phase 1 (until 2015): Initial preparation. This phase includes:
   a. **Prepare framework.** First steps in upgrading the electric grid where necessary; pilot fleets that can increase awareness, help advance technology, and drive down costs through learning, and initiate development of charging infrastructure; incentives for development of charging infrastructure; and introduction of national and international EV standards related to technology (plugs, etc.) and safety. The availability of a charging infrastructure is a basic requirement for deployment of EVs; standards are essential for ensuring compatibility of vehicles and charging across regions or national borders.
   b. **Increase renewable electricity generation.** For example, through fiscal incentives, public financing policies and regulations (see Chapter 5); grid stability.
   c. **Ensure development of a balanced grid.** Includes priority access to the grid for renewables, coordinated technical and institutional efforts, policies that support the development of a smart grid and active load management, and investment in long-distance transmission.

Pilot and other programs should start with urban areas, where population density is greatest and thus the costs are lowest per capita and goals can be realized more easily. However, charging infrastructure will also face more competition for space in urban areas, and a broader policy perspective that considers also how to reduce road traffic may also be required. Some countries will face greater challenges related to reach and strength of the electric

Continued

BOX 6.1 Policy Framework for the Coevolution of EVs and Renewable Electricity—cont'd

grid. Europe's grids, for example, can support higher penetrations than can those in North America and China. China, in particular, needs to focus first and foremost on expanding transmission capacity to integrate higher amounts of renewable generation.

2. Phase 2 (beyond 2015): Wide-scale deployment. This phase includes:

   a. **Help EVs to reach mass markets.** This can be done through use of tax or other incentives to address high initial cost, fuel subsidies, and granting of traffic privileges. In addition, enabling environmental policies such as information and marketing campaigns can be helpful, as can impact assessment of pilot projects and policy experiences in Phase 1.

   b. **Increase system integration.** Develop legal and technological frameworks for load management and vehicle-to-grid to enable higher use of EVs and renewable electricity.

General transportation sector policies can influence the deployment of EVs. These might include taxes on gasoline and diesel and putting a price on carbon through taxes or cap and trade, both of which make EVs more economically competitive. In addition, the coupling of renewable energy and EVs, while it could slow uptake of EVs, could establish a strong connection between the two.

Policy options include:

- **Hard coupling.** Electricity for EV charging is tied to the absolute additional share of renewable electricity in the power mix.
- **Emissions cap and trade.** Electricity production or vehicle deployment must meet specific emission targets, or additional electricity demand or vehicle deployment must be provided from carbon-neutral sources.
- **Grant manufacturers zero-emissions vehicle (ZEV) credits.** Vehicle manufacturers can count EV production (or sales) among their zero-emission vehicles if they finance new renewable electricity production.
- **Grid stabilization bonus.** Paid by system operator for plugged-in EVs that provide demand-side management or ancillary services.
- **Tax exemptions for charging.** EVs receive tax exemptions for charging with renewable electricity (most helpful where there are relatively high taxes on gasoline/diesel).
- **Reinvesting electricity tax from charging current.** Reinvesting revenues from electricity tax into new renewable capacity (most helpful where there are relatively high taxes on electricity).

Finally, one of the most important policy elements is for policy makers to have and clearly demonstrate a long-term perspective for development of related markets, industries, and infrastructure.

6.1. OVERVIEW—POLICIES FOR RENEWABLE TRANSPORTATION

A number of governments have established future targets for renewable energy's share of total primary or final energy supply, which indirectly affects the use of renewables in the transportation sector. The European Union has set a regional target for a 10% renewable share of transport energy by 2020, including biofuels, EVs, and electric trains (see Case Study 6). Most policies in the transport sector have been used to promote biofuels, and most of these policies have been in place for only a short time.

Biofuels Policies

Biofuels policies generally aim to promote domestic feedstock or biofuels production (e.g., with tax incentives or public finance for production facilities) or domestic consumption (e.g., through tax exemptions for blending or purchasing or blending mandates). The focus began with fiscal incentives and, in some cases, voluntary targets, but has been gradually shifting toward the use of regulatory policies (e.g., blending or production mandates) to advance biofuels production and use.

Tax incentives are the most commonly used form of *fiscal incentives* in this sector. They have been used along the entire value chain, but particularly in the form of excise tax credits or exemptions for producers, or reductions for biofuels at the pump to promote consumption. Other options employed include tax credits (e.g., per liter of fuel blended) for blenders of transportation fuel. As of early 2011, fuel tax exemptions and production subsidies were used in at least 19 countries, including 4 developing countries.

Meanwhile, the maturity of the sector is also illustrated by the expectation that the U.S. ethanol market will survive the expiration of the Volumetric Ethanol Excise Tax Credit (also known as the "Blenders' Credit") at the end of 2011, but only based on a mandated increasing blending level.

A few countries, including China and Indonesia, have provided direct support for biofuels through *public finance* (direct support of specific projects), with several other countries using government procurement to drive market growth. For example, the Thai government requires all of its fleets to use a gasoline-ethanol fuel blend.

Mandates typically address biofuel production or use levels, or the percentage required for blending with petroleum-based fuels (of energy or

CASE STUDY 6 EU Biofuels Directive

In 2009, the European Union (EU) approved two laws that affect the deployment of renewable energy fuels and technologies in the transportation sector: the Renewable Energy Directive and the Fuel Quality Directive. The Directives call for each EU member state to comply with a mandatory target of 10% sustainable transport by 2020, including both biofuels (with required innovation toward second generation fuels) and electricity for transport derived from renewable sources. Rules for fulfilling the Directives become stricter along the path toward 2020.

An earlier voluntary EU target, calling for biofuels to reach 5.75% of transportation fuels by 2010, brought about considerable growth in some countries but, according to early 2011 figures, only seven Member States (Germany, Sweden, France, Austria, Slovakia, Sweden, and Finland) achieved their targets. Moreover, the biofuels that qualified in national budgets were not subject to strong sustainability rules, such that their effect on greenhouse gas (GHG) reduction is expected to be limited.

In response, the European Commission (EC) established the obligatory 10% target. Although there is currently no clear penalty for nonachievement, the EC may take up additional measures upon receipt of interim reports, which must be submitted by Member States every two years.

In addition, the Directives clearly define what qualifies as renewable electricity and biofuels against this target. The renewables share in transportation equals the sum of the (liquid) biofuels part and the renewable sources share in the electricity input for EVs. Biofuels qualify only if they reduce GHG emissions by more than 35% compared to fossil fuels; this threshold will increase to a 60% emissions reduction by 2018.

Moreover, biofuels must comply with sustainability rules concerning land-use change and biodiversity, and these rules are still under development. In order to make these rules broad and applicable for the entire market, the EU links up and coordinates with international and national biofuels certification schemes. As of mid-2011, the EC had officially approved seven certification schemes for sugarcane- and soy-based biofuels, including some certification schemes that the industry applies to its own supply chain.

volume content). Renewable fuel/biofuel mandates had been enacted in at least 31 countries at the national level and 29 states/provinces as of early 2011. Most of these governments mandate that a specific share of ethanol or biodiesel be blended with gasoline or diesel, respectively, whereas the United States mandates a national blend level by volume. Since 2009, the EU has set a mandatory target of 10% sustainable transport by 2020 for

CASE STUDY 7 Brazil's experience with ethanol

Brazil's modern ethanol program began in 1974 when, in response to the global oil crisis, the government launched the PROALCOOL program with the dual aim of reducing the country's dependence on imported oil and supporting the sugar industry during years of low international sugar prices.

A national blending mandate combined with fuel tax exemptions and favorable government financing for ethanol production infrastructure (first attached to existing sugar mills, and later for autonomous distilleries) led to significant growth through the early 1980s. This growth was supported by parallel research to develop engines that could operate on pure ethanol, and by government efforts to develop a network for ethanol production and use, which was eventually turned over to the private sector.

But the industry struggled through the late 1980s and beyond, due to low international oil prices and a reduction in government incentives in 1986. As a result, the industry shifted from production of ethanol to sugar, ethanol supply shortages occurred in some regions, and consumers shifted away from ethanol and ethanol-only vehicles to cheaper gasoline. Purchases of vehicles that ran on hydrated ethanol plummeted from 88% of new vehicle sales in 1988 to 0.3% in 1996.

This all changed in 2003 when Brazilian auto manufacturers responded to government pressure, due to concerns about fluctuating ethanol supply and prices, by introducing flex-fuel motors. Flex-fuel cars can use ethanol, gasoline, or any blend of the two, giving motorists the flexibility to choose fuel based on price at the pump. Widely embraced by Brazilian drivers, flex-fuel cars witnessed strong demand growth and by 2008 they represented 89% of all new cars sold.

Parallel to development of the ethanol industry was the increasing use of bagasse (fibrous residue from sugarcane) for renewable heat and power generation in the sugarcane refining process. Early production was stimulated by government incentives. Now due to government regulations that have opened the electricity sector to private producers, mill owners can sell any excess electricity directly into the grid through contracts or auctions, which provides them with an additional source of income. During Brazil's 2010 harvesting season, sugarcane bagasse produced 18.5 TWh of electricity, including 8.8 TWh of excess that was sold into the electric grid.

For more than two decades, Brazil led the world in ethanol production; it was surpassed by the United States in 2006. By 2010, Brazil ranked second worldwide for both ethanol production and biomass power capacity. The only remaining government incentives were tax reductions for flex-fuel cars; ethanol subsidies were removed during the 1990s, and ethanol was cost-competitive with gasoline in Brazil without subsidies by 2004.

CASE STUDY 2   Brazil's experience with ethanol—cont'd

Key to early development in Brazil was the combination of mandates and incentives, along with development of necessary infrastructure. Other Latin American countries, including Guatemala and Argentina, enacted ethanol blending mandates during the 1980s in an effort to develop domestic markets. Brazil was the only one to support its mandate with a comprehensive package of fiscal incentives and government financing; it was also the only one to succeed at that time. As the market developed, vehicles that give consumers a sense of security of supply and choice of fuels were also critical as well as gasoline taxes that helped make ethanol cost competitive.

each individual Member State. This target includes both biofuels (with required innovation toward second generation fuels) and electricity for transport derived from renewable sources. Rules for fulfilling the Directives will become stricter along the path toward 2020 (see Case Study 7).

An advantage of obligations placed on fuel suppliers is the predictability (assuming enforcement) of market volumes to be reached in a given year. In addition, they do not result in lost tax revenue, a factor driving the shift from fuel tax exemptions to mandates in many countries.

Increasingly in recent years, governments have applied *sustainability and environmental standards* as an integral part of their biofuel support policies. This is most notable, perhaps, under the EU Renewable Energy Directive, which includes mandatory sustainability requirements, but is seen elsewhere as well. For example, the U.S. mandate includes a cellulosic biofuel requirement and applies GHG emissions thresholds for renewable fuels; India's National Biofuels Strategy mandates that biofuels be derived from nonedible feedstock grown on waste lands, degraded lands, or marginal lands; and Mexico prohibits the conversion of land from forest to agriculture for biofuels feedstock production. Brazil adopted a Social Fuel Seal as part of its biodiesel program requiring that specific requirements be met for producers to qualify for associated tax benefits. Quality standards like the American Society for Testing and Materials and the European Committee for Standardization are also of increasing importance to develop trust among consumers, vehicle manufacturers, and fuel suppliers, and can help simplify international trade of biofuels.

However, quality standards for the whole life cycle of biofuels are a complex matter, closely related to the complexity of life cycle analysis (LCAs). Comparison between the different fuels (also fossil fuels) should be based on such LCAs, which ask for proper assumptions, transparent system boundaries, and actual data.

Biofuels (and feedstocks) are internationally traded commodities, meaning that issues of competitiveness can affect the domestic situation. As a result, some governments have used *trade-related measures* to prevent exports or to protect domestic production via export/import tariffs or standards. The most common measures have been preferential trade agreements or import tariffs, used in combination with national support policies. Import tariffs, which have been included in biofuel mandates or used to prevent imported fuel from receiving domestic subsidies, shield national production and can help achieve production targets. However, this is also a potential cause of trade disputes.

Vehicle- and Infrastructure-Specific Policies

Policies to promote the development and use of FFV have included government procurement to stimulate investment in production, incentives for auto manufacturers that produce FFV, fiscal incentives for purchasers of such vehicles, and traffic privileges for flex-fuel drivers.

Brazil's modern ethanol program, which began in 1974, consisted of a national blending mandate combined with incentives and a parallel research program to develop ethanol-only engines. In later years, the focus shifted to support of FFVs and associated infrastructure.

In the past few years, policies have also emerged to advance the development and use of EVs and their associated infrastructure. Although the use of EVs does not imply that the electricity used will be renewable, such vehicles have the potential to play an important role in increasing the penetration of variable renewable technologies, and there is the option to link EVs and renewable electricity through policies in the future. (See Chapter 5 and Box 6.1 Policy Framework for the Coevolution of EVs and Renewable Electricity.)

As mentioned previously, several countries have announced targets that in combination would bring more than 200 million EVs to the world's roads by 2020. Some national and state/provincial governments subsidize the purchase of EVs through tax credits or exemptions, or through rebates. Pilot projects, mostly focused on EV deployment and upgrading the electric grid, are underway in a number of cities across Europe, the United States, and China. Some cities are developing charging infrastructure; and a few are now mandating the use of renewable power (supplied through green power purchasing) at these recharging stations. The granting of traffic privileges, such as free charging or access to certain city regions or traffic lanes, is another option currently in use to encourage EVs.

6.2. LESSONS LEARNED—POLICIES FOR RENEWABLE TRANSPORTATION

Fiscal incentives are more likely to stimulate the market without over-compensating producers, where they are adjusted regularly (and preferably automatically) based on relative fossil fuel and biofuel production costs. Tax exemptions have played an important role in the EU, and have been particularly successful in attracting investment in instances where fossil-fuel taxes have been high enough to account for the additional production costs associated with biofuels. Excise duty exemptions have successfully promoted biofuels investment in the early stages of market development in particular, as seen in the UK and Germany. However, losses of fuel tax revenue can be a drawback, and revenue losses increase as volumes rise. Further, removal of tax breaks can have unintended consequences, as demonstrated by Germany's experience with the phase out of tax exemptions for biodiesel, which led to a sharp decline in consumption. For more detailed information on the impacts of fiscal incentives and public finance, see Chapter 10.

Renewable fuel/biofuels *mandates* have been key drivers in the development of most biofuels industries. They have been determined to be the most effective policy option for increasing domestic biofuel production and consumption, and to be the most feasible from an institutional standpoint. Voluntary targets have also been known to advance production and use of biofuels in some countries, but their impact is quite limited compared with mandates, as seen in the EU (see Case Study 6).

However, mandates pose a number of challenges and thus need to be designed carefully and enacted with further requirements, such as sustainability criteria (including GHG emissions reductions, land use rights, etc.), or incentives. They have been criticized for associated indirect land use impacts (including a potential increase in GHG emissions), negative impacts on water quality, and for heightening global food insecurity. Further, they tend to promote the least-cost biofuels and are thus less effective in advancing specific biofuels types, including those from small-scale, distributed and regional production facilities. In Europe, for example, large-scale production facilities near waterways have tended to win out over regional small-scale distributed facilities, which do not have the same access to cheaper international imports of feedstock and fuels.

Most governments have enacted a combination of incentives, including mandates and other policy options, including the top producers of the United

States, the EU, and Brazil (see Case Studies 6 and 7). Recent developments in the EU demonstrate that countries with the highest shares of biofuels in transport fuel have systems combining mandates (which include penalties) with fiscal incentives (primarily tax exemptions). Meanwhile, it is important to design policy frameworks carefully to prevent the occurrence of unintended negative consequences, such as a net increase in total transport fuel consumption, which can result if fiscal incentives are combined with mandates.

Trade-related policies have also contributed to the effectiveness of domestic biofuel policies, such as production targets. But import tariffs, in particular, could jeopardize cost efficiency of domestic policies by preventing imports altogether or by limiting the amount of fuel (or feedstock) imported.

Vehicle- and Infrastructure-Specific Policies

Brazil's experience with biofuels over the years has demonstrated that the availability of FFVs can be of critical importance to the development of a strong market and industry. By allowing for greater use of biofuels in the vehicle fleet, FFVs increase potential consumption and, therefore, increase the attractiveness of investments in production. A combination of supply- and demand-side policies has been most successful in encouraging their production and use, as seen in Brazil. Sweden also was able to jump-start an FFV market through a variety of measures including public procurement (the city of Stockholm placed an initial order of 2,000 vehicles in 1998), investment grants for buyers of FFVs and exemptions from Stockholm congestion charges, energy and CO_2 tax exemptions for biofuels, and a mandate that large fueling stations provide high-blend biofuels. As a result, Sweden leads the EU for use and distribution of E85.

Analytical literature is limited regarding experience with policies to promote EVs, and particularly those driven by renewable energy specifically. However, lessons can be drawn from experiences to date with biofuels and, particularly, with policies to support the production and use of FFVs.

6.3. RECOMMENDATIONS FOR THE TRANSPORTATION SECTOR

Mandates have been found to be the most effective policy option for increasing domestic production and consumption of renewable fuels, but they must be paired with environmental and social criteria to ensure development is as sustainable as possible. That may be a complex challenge, however, and measurements for determining sustainability on the basis of

LCAs are still under development. At the same time, steps taken in the EU demonstrate that enough is known to make an important start in this direction. Further incentives might be desirable to assist in development of local or small-scale facilities or more advanced fuels and technologies, including preferential government purchasing.

As with the electricity sector, there are a number of design elements that may facilitate the effectiveness and efficiency of renewable transportation policies. These include:

- Establishing clear, consistent targets
- Implementing plans and programs to achieve targets
- Monitoring and reporting progress toward targets
- Designing and implementing evaluations to assess success
- Refining targets and programs based on evaluations or changing conditions.

Furthermore, support for renewable fuels—whether biofuels, electricity or otherwise—should be provided in conjunction with policies to increase efficiency of the national or regional vehicle fleet and investment in alternative and more efficient modes of transport (e.g., public transit, rail), as well as policies to reduce the need for transport altogether (such as encouraging telecommuting and making urban areas more pedestrian-friendly).

Government support is also critical to develop the necessary infrastructure (pipelines, transportation, charging stations, etc.), at least in the early stages of development, through public financing or fiscal incentives, and partnering with the private sector and municipalities to set up pilot cities for EV recharging stations, for example. Procurement of vehicles for government fleets can help create a stable market while increasing public awareness and acceptance. Standards related to fuel and vehicle safety, compatibility of systems, and so forth are also essential.

It is important to recognize that policies at different levels of government interact with one another; this is particularly true for biofuels, which are traded in the international marketplace. For example, international agreements can positively influence the implementation of national policies, and the success of national policies may be influenced or aided by regional programs that provide additional support for specific technologies. Likewise, international trade restrictions can hinder development. Subnational policies (e.g., state, province or municipality level) can provide supplemental support for national policies. Even where no national policies are in place, subnational policies can strongly influence the development of markets and supply chains, and in some instances they can pave the way for national level policies.

CHAPTER SEVEN

Heating and Cooling Policies

Contents

Heat services account for nearly half of global final energy demand—more than either electricity or transport. Their main uses are in buildings (for water and space heating) and industry (for process applications). Renewable heating and cooling (renewable energy H/C) have sometimes been called the "sleeping giants," because demand for such services is substantial, yet historically renewable energy policies have been focused primarily on renewable electricity or transport, thereby missing an opportunity to target the largest energy demand sector. As such, for policy makers wishing to push for additional renewable energy sources in the energy mix, renewable H/C represent low-hanging fruits that have often been overlooked.

Reducing H/C demand through energy efficiency measures, either in buildings or industry, works hand in hand with policy support to increase shares of renewable H/C in the energy mix. While this is true in other sectors as well, the potential for reducing energy demand in the heating and cooling sector remains very high. Efficient building designs, such as Germany's Passivhaus concept, significantly reduce energy demand through the use of uncommonly high levels of insulation, combined with "passive" heat from solar irradiation and tight air seals. In industry, combined heat and power (CHP) or heat recovery processes can harness and use energy that might otherwise be wasted.

This chapter provides a brief introduction to characteristics and challenges that are unique to renewable energy H/C technologies. It continues

Renewable Energy Action on Deployment
http://dx.doi.org/10.1016/B978-0-12-405519-3.00007-4

with examples of policy options used and others currently under development, and then provides analysis and lessons learned. This chapter concludes by presenting options available to policy makers in the near-term for increasing renewable H/C shares in the energy mix. It focuses on modern renewable energy applications in the buildings sector, although industrial examples are provided where relevant.

7.1. RENEWABLE H/C—WHY IT'S UNIQUE

Renewable energy sources used for heating and cooling include solar radiation, geothermal energy, and biomass in solid, liquid, or gaseous forms and derived from a variety of feedstocks (see Appendix B for a description of technologies). Renewable electricity is another option for meeting H/C demands with renewable sources (see Chapter 5). A majority of modern renewable energy heating technologies for buildings are mature and their development categorized as later-stage commercial, while renewable energy cooling technologies are typically at earlier stages of development.

Renewable H/C technologies present a significant challenge to policy makers for a variety of reasons. Systems are generally (though not always) small scale and distributed, and with no infrastructure to connect them together, rather than large scale and centralized applications that are connected by grids or pipelines. Depending on local infrastructure and circumstances, electricity generation and distribution occur in off-grid, decentralized systems, but in most cases they are centralized, and electricity is transmitted for up to hundreds of kilometers. Hot water or steam can also be transported from a central point to the end user, but the losses are generally larger than with power or fuel. So it is more challenging to transport heat or cold, resulting in a unique, multifaceted nature of the market that spans large-scale centralized district heating or cooling networks and small-scale decentralized systems of only a few kilowatts-thermal with onsite production.

While the demand side is fragmented with millions of building owners and developers producing and using heat, a diverse group of companies supply the market. Further, architects, installers, and engineers (often unwittingly) act as "gatekeepers" between supply and demand. As they are the ones who can open up (or close) doors to renewable energy integration, creating awareness, building capacities, and educating these groups will be essential.

The possible rate of increase of renewable energy in the total final heat energy demand of a region depends greatly on the existing infrastructure—those regions with fully functioning district heating systems, such as Sweden (see Case Study 8), may be most able to quickly integrate high shares of renewable energy. In contrast, regions where heat is produced and consumed on a highly distributed basis and where a large number of systems or fuels would need to be replaced will require more time for integration of high shares. Large-scale applications can provide entire cities with heating or cooling services, supply big industry, or even feed power plants (e.g., based on Organic Rankine Cycles). They are most often implemented in areas with substantial demand, such as densely populated areas or close to industry. Medium-scale applications can be installed in commercial buildings, or connected to district heating or cooling networks that supply a smaller area. Small-scale applications include small buildings, individual residences, or households. They are more common in areas that are less densely populated, where demand for heat is more widely distributed across space. The technologies available for these applications are as diverse as the applications themselves (see Table 7.1).

TABLE 7.1 Overview of renewable H/C technologies in centralized and decentralized applications

Renewable energy source	Select technology	Primary distribution method	
		Centralized	Decentralized
Solar energy	Low temperature solar thermal		X
	Solar cooling		X
Geothermal energy	Direct use applications	X	X
	Geothermal heat pumps		X
Bioenergy	Cookstoves		X
	Pellet-based domestic heating systems		X
	Small and large scale boilers	X	X
	Anaerobic digestion for biogas production	X	X
	Gaseous biofuels	X	

CASE STUDY 8 Sweden's Experience with Biomass for Heat and Power

Sweden has achieved a relatively high share of renewable energy (particularly biomass) in its heating sector, and increasingly for electricity generation as well. During 1980–2007, the biomass share in district heating (DH) production rose from 0 to 44% (90 PJ). As of 2009, the share of biomass in Sweden's total energy production exceeded that of oil—32–31%. Sweden is now one of the top producers of biomass power and heat in Europe.

Sweden's shift to a large share of biomass was facilitated by two previously existing infrastructure systems: (1) a large biomass resource and forestry industry, with well-established infrastructure and (2) its infrastructure for district heating, which provided 56% of the country's residential and service sector heat in 2008. Most of the expansion in Sweden's DH infrastructure occurred between 1965 and 1985, and it was driven primarily by high global oil prices and national energy taxes during the 1980s. It was also aided by the public's high acceptance for energy solutions as well as strong planning powers at the local level.

Growth in the use of biomass in the DH system took off after the government enacted a suite of policies in 1991. A gradually increasing carbon tax was adopted, accelerating the phase-out of oil for space heating in individual buildings and significantly increasing the use of biomass, which was exempt from the tax and became the least expensive fuel for DH systems. (The tax also drove the use of ground-source heat pumps and wood pellets for individual heat systems.) In addition, the supply chains and necessary infrastructure were already in place to immediately implement the biomass option.

As a result, the use of biomass for heat production in the DH system increased more than fourfold in six years—from 14 PJ in 1990 to 61 PJ in 1996—and rose again to 71 PJ by 2002 and 101 PJ by 2009 (about 71% of total district heat input;) see Figure 7.1). There was little resulting impact on biomass power capacity in the early 1990s because the carbon tax focused on heat.

In 1996, the electricity market was liberalized and the government enacted investment subsidies for biomass-based CHP plants, small-scale hydropower, and wind. These subsidies are believed to be responsible for a short-term increase in biomass capacity; the growth was short-lived, perhaps because of the short subsidy period (through 2002). Carbon taxes were increased again in 2001, resulting in significant increases in output, but not capacity.

Driven greatly by a goal to replace nuclear power, in 2003 the Swedish government introduced an electricity quota obligation with tradable renewable electricity certificates (TRECs), which sets a mandated quota for generation from all producers except electricity-intensive industries. Renewable energy producers receive income from the sale of electricity (at the market price) and from the sale of TRECs, which are traded separately on the market. The regulation has been revised and extended several times, with the obligation rising from an initial 10 TWh above 2002 levels by

CASE STUDY 8 Sweden's Experience with Biomass for Heat and Power—cont'd

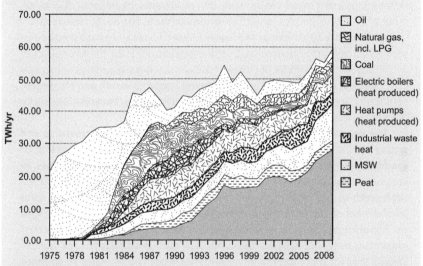

FIGURE 7.1 Energy input for district heating in Sweden, 1970–2009. (*Mitchell et al., 2011; Swedish Energy Agency, 2010.*) For color version of this figure, the reader is referred to the online version of this book.

2010 to 25 TWh above 2002 levels required by 2020 (as of 2009). Tax incentives and investment subsidies, for wind power plants in particular, were also implemented.

Renewable electricity production increased significantly in response to the scheme, particularly biomass-based power production in CHP plants; as a result, heat production for the DH system also increased markedly. The TREC scheme has provided a stable investment environment and has achieved its targets.

However, while other renewables have expanded in the electricity sector they have not done as well as biomass. In response, the TREC scheme has been criticized for overcompensating biomass and providing prices that are too low to generate investment in more expensive renewable energy technologies such as solar photovoltaics (PVs). In addition, the quota system covered some existing capacity, leading to "free riding" and windfall profits, and some experts posit that that carbon tax was at least as important as the quota obligation for increasing output.

To further promote renewables beyond biomass, a support scheme was introduced in 2009 that offered support to solar PV and hybrid PV/heating systems, covering up to 40% of the investment cost. An installation can receive both investment support and income from sales of electricity certificates. Wind power investments were restricted by the short time frame of the policy in early years, but picked up after the scheme was extended in 2006 and a 15 year support period was adopted. Permitting and planning procedural hurdles have since

Continued

CASE STUDY 8 Sweden's Experience with Biomass for Heat
and Power—cont'd

become important barriers to investment in wind power, highlighting the importance of "enabling" policies to overcome non-economic barriers to renewables.

As of January 2012 Norway and Sweden joined forces and established a common electricity certificate market. Over the period until 2020, the two countries
aim to increase their production of electricity from renewable energy sources by
26.4 TWh. The joint market will permit trading in both Swedish and Norwegian
certificates, and it will be possible to receive certificates for renewable electricity
production in either country.

*The key to Sweden's success, in addition to existing infrastructures and
supply chains and low-cost biomass resources, has been a combination of
gradually increasing energy and CO2 taxes on one hand, and policies to
support renewable energy on the other (fiscal incentives for heat, and fiscal incentives and regulations for electricity). Also important has been local
involvement in the planning process. Difficulties with wind power highlight
the importance of educating and engaging the local population in a way that
maximizes potential benefits for all. Further, the Swedish experience demonstrates that for more expensive renewable energy technologies to advance
under a quota system it is helpful to provide additional support.*

Some Specific Barriers for Renewable H/C

For some renewable H/C technologies, system components have not yet
been developed that would allow for large-scale deployment. For example,
in larger solar thermal systems used for industrial process heat, fundamental
R&D is still required on system components such as storage materials.

Environmental regulations restricting the reinjection of groundwater
can also act as a barrier to direct-use geothermal installations. Barriers specific to ground-source heat pumps include: (1) cost and difficulty of evaluating the suitability of individual installation sites, (2) installation-specific
design and engineering requirements, and (3) space requirements for ground
coupling in densely built areas.

Barriers specific to biomass heat are strongly related to the availability
of feedstock, as well as the supply chain infrastructure for processing those
feedstocks. In areas where there is either a lack of resource base, or where
the infrastructure for processing the feedstocks has not yet been developed,
increased deployment of biomass heat will face additional challenges.

There are also challenges with regard to linking a variety of renewable energy systems, for example, finding the most efficient way to combine solar thermal and ground-source heat pump systems in different situations and climates. Another typical barrier to renewable energy H/C installations in buildings is the question of "split incentives"; for instance, where the return on investment will go to the building occupants or renters, building owners have little incentive to invest in efficiency improvements and renewable energy H/C technologies.

7.2. OVERVIEW—POLICIES FOR RENEWABLE H/C

There is limited experience with policies to advance the use of renewable H/C technologies. This section provides an in-depth look into the types of policy instruments that have been implemented in support of renewable energy H/C technologies, highlighting important design features and considerations where possible. As with the other end-use sectors, determining which barriers most hinder renewable H/C technologies in the targeted country or region at the early stages of policy design may help to steer the development of the most efficient and effective package of support policies.

Renewable H/C policy instruments are structured into the following categories: fiscal incentives, public finance, and regulations. Policy instruments in each of these categories may be applied to support renewable energy H/C technologies, although the extent to which each policy type may be applied to different renewable energy H/C applications (e.g., large-scale centralized vs. small-scale decentralized) can vary considerably. Some instruments may be more challenging to implement on a small scale and decentralized basis (e.g., FITs for heat) and are therefore more readily suited to large-scale or industrial applications. Others, such as use-obligations provide an immediate incentive for individual buildings, but may or may not support larger scale applications such as integration of renewable energy into district heating systems. In both cases, design elements can be introduced to facilitate application of the policy to broader H/C scales and installations if desired. For some policy types, including grants or loans, there are design elements of the policies themselves that may be better suited to large- or small-scale applications. In Section 7.3, a subsection at the conclusion of each policy category briefly discusses the applicability of policy instruments and their design features to renewable energy H/C applications of varying scale.

Fiscal Incentives

Fiscal incentives, including grants, rebates, tax credits, or tax reductions, have been employed more than any other political instrument in support of H/C technologies. They can be targeted to either installation costs of a heating and cooling plant (investment incentives) or to the actual generation of heat (production incentives).

Grants supporting renewable energy H/C technologies can target either the investment costs of the installations themselves (capital grants) or the heat generated from a given installation on, for example, a currency/GJ heat-generated basis (operation grants).

Tax-based incentive systems can include tax reductions or exemptions, such as from sales, property or value-added taxes, thereby reducing the total cost of investment. Tax credits (production or investment) give renewable energy H/C system owners an annual income tax credit, meaning that investments can be fully or partially deducted from tax obligations or income.

Indirect incentives for renewable energy H/C technologies are also possible via exemptions from eco-taxes, carbon taxes, or other levies on fossil-fuel–based energy. Without providing incentives to renewable energy H/C specifically, such exemptions help to level the playing field for renewable energy H/C technologies, making them more cost competitive and thereby a more attractive option.

Public Finance

Public finance policies are less common than fiscal incentives, although they have been implemented in some instances. They include policies such as low-interest loans and guarantees. Often the amount available for loan or guarantee depends on the scale of the project.

Regulations

Regulations include obligations, mandates, and recently also premium payments, or feed-in tariffs (FITs) for heat. Most regulations in place have been use-obligations for solar thermal technologies, although some examples exist for other technologies, such as biomass heat in Denmark. Alternatives to use obligations are rare, although an FIT for heat—the first of its kind—was recently implemented in the UK. Although there is limited experience with regulations for renewable energy H/C technologies, there is growing interest in this type of policy support due to its independence from public budgets.

Use obligations require installation of a renewable energy H/C system in new buildings or in buildings undergoing substantial renovation. They inherently target decentralized heating or cooling applications in commercial or residential buildings.

The first use obligation for solar thermal was implemented in Israel in 1980, and was the only one of its kind for nearly two decades. Israel's early solar thermal obligation applied to all new buildings except hospitals and those used for industrial or trade purposes; it also excluded buildings higher than 27 meter. Beginning in 2000, similar policies were implemented on a municipal level in a number of European countries, and have since been implemented on a national scale in Spain, Germany, India, South Korea, and Uruguay.

FITs are widely known instruments for supporting renewable power generation (see Chapter 5), but there is very little practical experience with these policies in H/C markets. FITs for heat establish a purchase obligation with fixed reimbursement rates for the generation of renewable heating or cooling, similar to FITs for electricity. Owners or operators of H/C installations receive a fixed price per kWh corresponding to the amount of renewable heat that they produce. The level is determined by the government and set by law.

In March 2011, the UK introduced the world's first FIT for heat in its Renewable Heat Incentive (RHI) policy; in March 2012, the Netherlands followed the example with a feed-in premium for renewable heat. The British RHI will be implemented in two phases: the first targeting large-scale industrial heating, business, and community heating projects and the second offering long-term tariff support for the domestic sector. RHI applies to solid and gaseous biomass, solar thermal, ground and water source heat-pumps, onsite biogas, deep geothermal, heat energy derived from organic waste, and injection of biomethane into the grid, as well as renewable energy generated heat used for cooling purposes. Heat generation will be measured via meters, whose installation is required at the point of generation and (where appropriate) at the point of use. Standards will be applied to assure that meters are of a certain quality and reliability. Program participants are required to submit schematic diagrams of their installations, including the locations of all meters, as part of the quality assurance process. Payments will be made on a quarterly basis over a 20 year period. If successful, this program could serve as a future model for national and subnational policies alike.

FITs for heat distribute additional costs among fuel consumers according to the "polluter pays" principle, and they enable a reliable return on investment, thereby providing a secure calculation base independent of public budgets. Therefore, they are considered to be advantageous when compared with fiscal incentives or use obligations.

7.3. LESSONS LEARNED: POLICIES FOR HEATING AND COOLING

The number of policies in place to support renewable H/C technologies worldwide has been increasing since 1990. This section reviews current policy experiences and provides case studies with policy examples on the subnational (state/provincial), national, and international levels.

Despite the compartmentalization here, it is important to consider the interplay among the different levels of policies. International agreements can positively influence the implementation of national policies, as seen in Europe, where inclusion of the heat sector in the 2009 Renewable Energy Directive has driven some Member States to implement strong national policies; as a result, renewable H/C in the region is expected to increase substantially by 2020 (see Case Study 9).

The success of national policies may be influenced or aided by programs at higher (international) or lower (state/provincial/local) levels of governance that provide additional support for specific technologies. Renewable H/C support policies implemented on the national level have typically been technology specific and with varying degrees of success.

Even where no national policies are in place, subnational (e.g., state, province or municipality level) policies can strongly influence the development of markets and supply chains. As with the electricity and transport sectors, subnational policies can act as important supplemental support for national policies, or can build renewable energy H/C markets in the absence of a strong national policy; in some cases they can even pave the way to national level policies. Canadian provinces and U.S. states, for example, have played significant roles in advancing renewable technologies in both of these countries; their renewable heating policies have provided important additional support to national policies and have also promoted development of strong local markets (see Case Study 10). Because of the limited experience with policies to support renewable cooling, the majority of the focus of this section is on experiences with renewable heating policies.

CASE STUDY 9 The Influence of an International Directive on Renewable Energy H/C

Growth of renewable energy in Europe's H/C sector has been modest in comparison with growth in the renewable energy electricity and transport sectors (RE-Shaping, 2011); this may be due to the lack of support frameworks at the European level and at the national level in most European Union (EU) countries. Domestic, decentralized biomass heat systems (such as small-scale wood boilers) dominate the renewable heat market in Europe, with biomass in CHP plants and solar thermal systems playing a lesser role.

In April 2009, the EU adopted a directive that established a binding EU-wide target for a 20% share of renewable energies in the energy mix by 2020 (European Parliament, 2009). Prior to implementation of the directive, renewable-based heat generation in Europe had increased from 452 TWh (38.87 Mtoe) in 1990 to 770 TWh (66.22 Mtoe) in 2008: an average annual growth rate of 3%. Although it is not yet possible to reach a conclusion about the influence of the 2009 EU Directive, the National Renewable Energy Action Plans (NREAP) of the Member States provide insights into the possible development of the renewable heating markets (assuming all Member States reach their targets as outlined).

Renewable H/C in the 27 Member States is expected to increase to 111.5 Mtoe (1,296.7 TWh) by 2020, nearly doubling the current contribution. Biomass heat is expected to continue to dominate the renewable energy heating market, although shares of solar thermal, deep geothermal heat, and heat pumps also increase to reach 6.3 Mtoe (73.3 TWh), 2.6 Mtoe (30.2 TWh), and 12.1 Mtoe (140.7 TWh) of generation, respectively, in 2020 (see Figure 7.2).

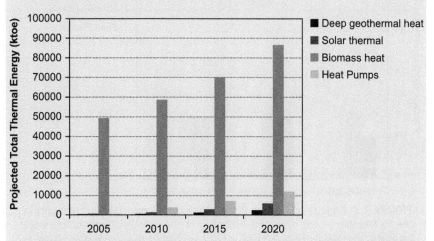

FIGURE 7.2 Projected total renewable energy heating and cooling energy in the 27 Member States of the EU according to the NREAPs. *(Data sourced from Beurskens and Hekkenberg, 2011.)* For color version of this figure, the reader is referred to the online version of this book.

CASE STUDY 9 The Influence of an International Directive on Renewable Energy H/C—cont'd

The expected average annual growth rates of the different renewable energy H/C technologies show an early increase in heat pump deployment (largely due to the individual country activities in Italy, France, Belgium, and Finland) and relatively constant growth rates for solar thermal and geothermal heat technologies (at 10–16% annually). The average annual growth rate of biomass remains near current levels, increasing to 4.5% in 2020 (see Figure 7.3).

Although some EU Member States (particularly Germany and Sweden) have implemented strong national policies to support the development of renewable energy H/C markets, the average annual growth rate across the region has remained around 3%. Assuming that all targets for renewable energy deployment under the new EU Directive are met, the average annual growth rate in 2020 across technologies is expected to increase to 11.4%.

The European example demonstrates that an international policy has the potential to increase the shares of renewable energy H/C across a multinational area. Strong renewable energy H/C markets are currently concentrated in the few Member States with national supporting policies in place. If targets for the 2009 Directive are met, markets will have developed even in areas where currently only nascent or nonexistent markets are in place—helping to support a wide, stable deployment base of renewable energy H/C technologies across countries.

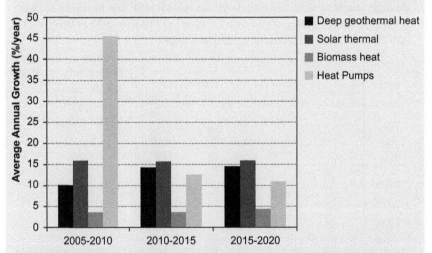

FIGURE 7.3 Average annual growth of renewable energy H/C technologies in the 27 Member States of the EU according to the NREAPs. *(Data sourced from Beurskens and Hekkenberg, 2011.)* For color version of this figure, the reader is referred to the online version of this book.

CASE STUDY 10   Provincial/State Programs in Canada and the United States

As demonstrated by provinces and states in Canada and the United States, respectively, subnational policies can provide additional backing to national policies and can promote localized development of renewable energy H/C markets.

In Canada, the national ecoENERGY for Renewable Heat Program (2007–2011) was supported by complementary provincial programs in British Columbia, Ontario, and Saskatchewan. It offered matching grants to the industrial/commercial/institutional sectors for installation of solar thermal installations used for space heating and cooling and for water heating. Ontario has also experimented with rental programs for solar hot water systems, supported in part by the ecoENERGY program, to provide an alternative for addressing the barrier of high up-front costs. At the end of 2010, Canada had installed solar thermal capacity equivalent to 712 MW_{th}, driven in great part by these provincial programs.

In 2005, the United States implemented a 30% federal tax credit (up to USD 2,000) for the purchase and installation of residential solar hot water heating systems, and 10% for geothermal direct-use installations and heat pumps. These incentives were expanded in early 2009 as part of the federal stimulus program. However, most of the country's support for renewable energy H/C has been implemented on the state level. As a result, H/C technology markets are dispersed, reflecting the areas of regional support for the different technologies. For example, strong state support for solar thermal technologies has facilitated the development of solar thermal markets in Hawaii and California, where the majority of U.S. capacity has been concentrated.

Hawaii has supported solar thermal installations since 1959, when the state began offering specialty licenses for solar contractors, including those specializing in solar heating and cooling technologies. Since 1976, tax credits have been available for 35% of the cost of equipment and installation of a solar thermal system (with interim amendments) and, beginning in 1996, a one-time rebate of USD 750 was offered for solar hot water installations. Some local utilities have also offered zero-interest loans and rebate programs. As of January 2010, Hawaii implemented the United States' first use obligation, requiring the installation of solar hot water heaters on all new construction of single-family homes (with a few exceptions).

Hawaii's long-term, reliable and multifaceted "carrot-and-stick" approach (based largely on fiscal incentives and now also including regulations) has combined with high conventional energy prices to make Hawaii the center of the U.S. solar thermal market for many years.

Some countries (e.g., Germany and Sweden) have witnessed strong growth in renewable heating markets in response to national policies. Sweden has increased the share of biomass in its heat supply substantially since the 1980s, due to a suite of support policies. In addition to existing infrastructures and supply chains and low-cost biomass resources, Sweden's success has resulted from a combination of gradually increasing energy and CO_2 taxes on one hand, and fiscal incentives to support renewable heat on the other (see Case study 8)

However, other countries have had very little market growth despite policy support. In Japan, for example, the decline of the solar thermal market since the 1990s, despite the presence of technology-specific support under the New Energy Law (in place since 1997), shows the importance of influential factors outside of policy design elements. Over the years, the decrease in the price of conventional fuels and a strong local currency has overshadowed the direct policy support for solar thermal technologies in Japan.

Although policies have been critical in increasing deployment levels of renewable H/C technologies in place today, there are also some markets that have developed independently of political support. In China, for example, the solar thermal market has been driven by consumer demand for affordable, reliable heating systems in an environment where there was a lack of reliable conventional heating options and supplies, combined with a strong solar resource and comparatively inexpensive installation costs (approximately USD 0.25/W compared with USD 2/W in most other countries). By the end of 2008, China had more installed solar thermal capacity than any other country worldwide, with a total of 87.5 GW_{th} and representing 66% of the world market for flat plate and evacuated tube collectors. Reflecting the manufacturing infrastructure that has been built to supply the demand for solar thermal technologies in China, exports of solar hot water heaters have also increased substantially with a nearly sixfold increase between 2001 and 2007.

Fiscal Incentives

In general, fiscal incentives that focus support on the generation of heat, rather than investment in systems or subsidies to producers, may be preferable as emphasis is placed on plant performance, thereby encouraging better quality installations. (For a more detailed analysis of fiscal incentives and public finance policies, see Chapter 5.) Where scale and application determine that it is advantageous to place incentives on installation costs rather than generation, it is important to link support to standards related to installation and performance.

India's experience with its National Program on Improved Chulhas (NPIC) provides an example of how subsidies for producers can inhibit development and uptake of technologies if they are not combined with performance criteria. The NPIC offered subsidies to improved cookstove producers starting in 1983, with the aim of promoting efficient use of biomass fuels for cooking, but no quality standards were implemented alongside the subsidy. Because producers had little incentive to consider consumer preferences, efficiency was not greatly improved and the quality of stoves produced was poor. Although 80 different improved stove models were developed over the course of the program, public perception of the technology was damaged; in 2002, the program was discontinued and responsibility for making further improvements was passed to the states.

In addition to offering incentives according to the amount of heat or coolness generated (rather than to producers or based on capacity or as a percentage of cost), several possible design features are applicable to fiscal incentives that may help to assure quality installations. For example:

- Offer an additional incentive for higher quality equipment
- Request and review technical drawings from installers before incentive funds are granted
- Request an energy audit certificate as part of the application for incentive funds

Ensuring the generation of heating or cooling service by means of quality systems helps to develop consumer confidence in the technology and the market and therefore provides positive reinforcement to the goals of the policymaker.

Grants have been successful in increasing the shares of renewable energy heat in a number of countries, including Germany (solar thermal), Austria (biomass-fueled heating systems), and Switzerland (heat pumps). But consistent and long-term application is important for ensuring investor confidence and a stable market.

As a part of Germany's renewable energy policies portfolio (see Case Study 12) the Market Incentive Program (Marktanreizprogramm; MAP)—targeting solar thermal systems, a variety of biomass heating systems and heat pumps—is one of the most cited examples of successful grant policies for renewable heat. Between 1999 and the end of 2007, the MAP supported 650,000 installations with a total investment volume of USD

8.5 billion (€6.5 billion). Because of its long-term commitment, the MAP is well known to professionals in the industry who typically advise their customers of its availability. However, over its lifetime the MAP has been amended several times, changing the level of support for different technologies, eligibility requirements, and application procedures; these amendments have been accompanied by fluctuations in application rates, inferring concerns about reliability of the fund. While it has been undeniably successful in promoting renewable heating technologies due to its long-term support, experience has shown the importance of ensuring consistent, reliable signals over time. In general, fluctuations in annual support levels have undermined investor confidence, which may in turn inhibit the success of the policy.

Further, it is important to carefully consider possible implications for the public budget when designing grant-based incentive policies. The dependence of grants or other fiscal incentives on public budgets may be considered a disadvantage because such incentives are indirectly dependent on the existing political agenda (or economic circumstances) at any point in time. Therefore, grants and other fiscal incentives cannot necessarily function as permanent, reliable forms of policy support.

It has been suggested that *tax-based policies* such as tax rebates may be advantageous over grants or subsidies (requiring pre-approval), as owners are not necessarily obligated to wait for approval processing before beginning installation of the system. Rather, administrative requirements can be minimized with ex-post processing. For example, the French 2005 Finance Law established a procedure by which owners of solar thermal or biomass-based heat installations could recover costs via an income tax declaration, simplifying the process.

Fiscal Incentives for Large- and Medium-Scale Applications

In many areas, one clear challenge to implementing renewable H/C on a large scale is a lack of necessary infrastructure such as district heating or cooling networks. While fiscal incentives may be directly applied to renewable energy H/C technologies, supporting the development of large-scale, capital-intensive infrastructure into which renewable energy-generated heating or cooling can be readily fed is also an important option for consideration.

Specific design features of fiscal incentives may be better tailored to large- or medium-scale applications. For example, including meters or monitoring technologies in large-scale, centralized applications is more common and less of a financial burden as a share of the up-front investment.

As a result, fiscal incentives targeting the generation of renewable-based heating or cooling may be better suited to large- or medium-scale technologies and applications.

Fiscal Incentives for Small-Scale Applications

Due to the challenges of monitoring renewable H/C generation on a smaller scale, fiscal incentives based on installation costs (vs. the actual generation of heat) may be best suited for small-scale applications. While appropriate steps should be taken to assure quality, fiscal incentives based on installation costs address up-front costs, which are often considered a significant barrier to residential or individual building systems, thereby effectively supporting decentralized applications.

Public Finance Policies

To be effective, *loans* for renewable energy H/C projects should be long term (at least 10 years), have a low interest rate, and have low hassle and administrative fees. Such instruments are successful largely where the underlying value proposition of the technology is high; where the economics of a project are less attractive, public finance instruments are typically less successful.

Examples of low-interest loans for renewable heating exist in Austria and several U.S. states. In France, the Crediting System in Favor of Energy Management guarantees up to 70% of the total investment through bank loans requested for renewable energy (including heating) and energy-efficiency projects.

As is the case for renewable electricity, public finance instruments may be applied to large-, medium-, and small-scale renewable H/C applications. It is most often the design features of the policy that determine if a low-interest loan is most attractive for large-scale utility installations or for small-scale residential applications. For example, application fees that are above a certain threshold may make residential projects unattractive, whereas such fees in larger scale projects are comparatively more affordable.

Regulations

Experience with *use obligations* shows the importance of including quality assurance measures in the early design of the policy. Spain offers an example of a country that has seen great success with obligations for the installation of solar thermal systems in combination with quality assurance policies such as technical requirements (see Case Study 11). Such quality measures can include:

- Quality parameters for the products and system configurations
- Installation works guarantee and after-sale maintenance
- Third-party monitoring of a sample of the systems installed
- A clear sanctioning regime

CASE STUDY 11 Spain's Solar Thermal Use Obligations

In 2006, Spain implemented its Technical Building Code (Código Técnico de la Edificación; CTE), obliging owners of all new buildings and those undergoing renovation to provide 30–70% of domestic hot water demand with solar thermal energy. The national Spanish obligation was implemented following several successful municipal solar ordinances of similar nature in Barcelona (2000), Seville and Pamplona (2002), and Madrid (2003). The passing of the Spanish CTE was enough to trigger substantial investment and expansion in solar thermal production lines across Europe.

The national government took advantage of lessons learned from similar policies implemented at the municipal level regarding the importance of including quality assurances for installation and requirements for ongoing maintenance. In order to assure quality installations, the CTE

...defines a number of technical requirements on the components, design and installation of the solar thermal system, including sections on the solar collector and its components, the working fluid, the storage systems, the hydraulic circuit, the controllers and the conventional auxiliary system.

Such measures assure consumer confidence in the technology and the ongoing production of renewable heat. Although there were quality issues with installations in the beginning (and despite the economic downturn in the building sector), the CTE and municipal ordinances have been the main drivers for the Spanish solar thermal market.

Without such quality measures, builders may be more likely to install cheaper, poorer quality products simply to fulfill their obligation, potentially hindering or even actively dissuading public acceptance of the technology. Emphasis should be placed on the continued generation of renewable heating, rather than simply the plant installation.

As mentioned above, FITs are generally considered advantageous when compared with fiscal incentives or used obligations because they distribute additional costs among consumers according to their consumption levels and enable a reliable return on investment. As of publication, however, there is very little experience with this policy type for the heating and cooling sector.

Regulations for Large- and Medium-Scale Applications

Use-obligations targeted to individual buildings provide little incentive for increasing the share of renewable energy sources in district heating systems (except where network operators and heat suppliers are included in the obligation) or the construction or use of these systems. Specific design elements or specifications must be incorporated into the policy if it is a priority for policy makers to target large-scale or industrial applications with a use obligation.

However, it might be possible to implement a variation in use obligation policies that would allow building owners the option to pay a substitute levy instead of directly fulfilling the use obligation. Resulting levy funds could be used to promote the construction of larger scale systems such as district H/C networks. Such a variation has not yet been implemented in practice, but may offer an alternative for policy makers wishing to target both centralized and decentralized systems.

As FITs are based on payments per unit of heat energy produced (kWh_{th}), meters are needed to measure the amount of heat generated. Such a requirement can be onerous for small-scale applications, but generation meters are often common practice for systems over 300 kW, and meter installation costs represent a small share of overall costs for larger-scale applications (DECC, 2011). FITs can also be applied to biomethane that is injected into natural gas grids for use in combined heat and power plants, for example.

Regulations for Small-Scale Applications

Because *use-obligations* inherently target decentralized heating or cooling applications in commercial or residential buildings, they are well suited to support small-scale renewable H/C applications.

Implementing FITs in the heat market may be more challenging than in the electricity market, particularly for distributed, small-scale heating and cooling applications. There are two options for policy makers interested in FITs for small-scale applications.

- Require the implementation of meters on all small-scale generating units, noting that the cost of such an option has yet to be quantified.

In practice, residential applications rarely have metering or monitoring instrumentation. The costliness of implementing additional equipment may or may not impede the attractiveness of the incentive.

- Establish different requirements for different sized systems, as was proposed by the tentative German Renewable Energies Heating Act. Small systems could be given a simple rated output level, removing the requirement for regular data collection and monitoring.

7.4. RECOMMENDATIONS FOR THE H/C SECTOR

Substantial potential exists to increase the share of renewable energy in heating and cooling supplies worldwide. No matter the specific type of policy instrument implemented in support of renewable energy H/C technologies, there are a number of design elements that may facilitate their effectiveness and efficiency. These include the following:

- Establishing clear, consistent targets
- Implementing plans and programs to achieve targets
- Monitoring and reporting progress toward targets
- Designing and implementing evaluations to assess success
- Refining targets and programs based on evaluations or changing conditions.

Moreover, there are a number of steps policy makers can pursue immediately to begin scaling up the use of renewable H/C technologies. Perhaps most important is to enact policies to advance energy efficiency (e.g., through incentives, building standards, or other measures) in parallel with policies to support renewable H/C. The potential for reducing energy demand in the H/C sector is very high, and efficiency improvements can significantly increase comfort levels while reducing costs associated with a shift to renewable H/C.

Prior to implementing policy support for renewable energy in this sector, it is important to gain a clear understanding of the following:

- The structure of the H/C markets in place in the country or region (whether centralized or decentralized)
- The status of existing renewable H/C supply chains— whether the infrastructure has already been developed to supply the technologies (and feedstocks, in the case of bioenergy)
- Policies already in place on relevant, complementary levels (e.g., municipal, state/provincial, or national) that may positively (e.g., additional incentives) or negatively (e.g., planning regulations/restrictions) influence the success of additional renewable H/C policies.

It is then necessary to determine how scale-up of renewable H/C is to be prioritized. In other words, is it a priority to implement new infrastructure such as H/C networks that may help facilitate large-scale increases in renewable H/C use? Or is it a priority to expand the use of renewable H/C technologies within the existing infrastructure?

Another important issue here is building up the capacity and competences of the workforce (architects, installers, building and construction workers, etc.). As the infrastructure for heating and cooling in many countries is very decentralized and systems are small, changing the infrastructure and the systems will require quite an effort of the workforce. Communication and education are essential.

There is no one-size-fits-all policy package. Once the conditions and relevant factors involved in renewable H/C deployment in a specific region are identified, and priorities and the available short- and long-term public budget are clearly delineated, it is possible to design a policy package that is best suited for promoting deployment of renewable H/C in a particular location, based on the lessons identified above.

CHAPTER EIGHT

Energy Systems Change—Policies for the Transition

Contents

The successful transition to a sustainable energy future will require both the integration of a broader range of renewable technologies into the energy system, and a significant increase in the overall share of renewable energy in overall supply. To achieve these aims, a greater level of systems thinking will be needed. This will require a more holistic view of policy making that sees renewables, energy efficiency, and energy generally (across all end-use sectors), in addition to agriculture, forestry, education and training, water, construction, urban planning, and so forth, as interconnected sectors. Efforts will need to be made, and coordinated, at all levels of government and society, as well as across state and national borders.

Increased attention and resources will need to be focused toward flexible technologies and infrastructures such as smart grids and transmission capacity. As mentioned in Chapter 6, coevolution of renewable energy and electric vehicles could play an important role in advancing the transition in the transportation sector while also helping to integrate high penetrations of variable renewables.

A shift from fossil fuels to renewable energy will need to go hand-in-hand with energy-efficiency improvements on the supply (energy production) and demand sides if countries are to achieve sustainable energy systems. The supply side covers all energy sources and carriers and their associated delivery networks, including pipelines and electricity transmission and distribution grids. Policies on the demand side cover the range of end-use sectors from residential and commercial buildings (space and water heating and cooling, cooking, lighting, electronics, and other electricity uses), to industry (process heat, motive power), to transportation.

Policies that advance energy efficiency can complement renewable energy policies by taking advantage of the many synergies between

Renewable Energy Action on Deployment
http://dx.doi.org/10.1016/B978-0-12-405519-3.00008-6

137

renewable energy and energy efficiency. For example, improvements in energy efficiency enable renewables to more rapidly and quickly achieve a larger share of total energy supply. Further, wherever renewables displace thermal processes (using fuel or nuclear power), the amount of primary energy required can be reduced significantly. The use of distributed renewables can reduce transmission and transportation losses, again reducing primary energy requirements. Direct use of solar energy for passive lighting and heating can provide required energy services without the use of energy conversion technologies. Further, the use of smart-meters and responsive demand and time-of-use pricing can shift demand load in a way that benefits system operation and matches demand to the supply of renewable energy. Ultimately, what people need and want are energy services, such as lighting, heat, and mobility. There are many options as to how to supply these services; for example, lighting can be provided by daylight.

Germany (Case Study 12), Denmark (Case Study 13), and Upper Austria (see Case Study 14) provide examples of countries and regions that have already achieved high and significant shares of renewable energy through a variety of means and that have started down the road toward transformation.

CASE STUDY 12 Germany's Path toward Energy System Transformation

Germany has devoted significant resources to renewable energy technology development and deployment since the 1970s. Technologies advanced through government R&D efforts, but faced a hostile environment in the 1980s; declining oil prices made it difficult for renewables to compete in the market, and the large utilities dominating the electricity supply system were opposed to small, decentralized forms of generation. Despite the odds, and in response to the introduction of more favorable policies, Germany's renewable share of primary energy supply increased from 1.3% in 1990 to 9.4% in 2010, with most of this increase in the previous five years.

In 1989, the government established a small subsidy for the first 100 MW of wind power installed in Germany in exchange for a requirement that recipients reported on turbine performance. Germany's first feed-in tariff (FIT) was enacted the following year.

Until 2000, the FIT guaranteed producers access to the grid and the sale of power at a specified percentage of the retail rate. It was revised in 2000, due in great part to developers' concerns about falling electricity prices following deregulation of the German market, which reduced payment to renewable generators. The new FIT (under the Renewable Energy Sources Act of 2000; EEG,

CASE STUDY 12 Germany's Path toward Energy System Transformation—cont'd

2000) covered more technologies and introduced cost-based tariffs that varied by technology. Tariffs were guaranteed to all renewable generators for at least 20 years, with remuneration for new plants decreasing at a predetermined rate. Grid operators and electricity suppliers are obligated to provide priority feed-in and dispatch. The EEG was revised again in 2004, reducing tariffs for wind power (and eliminating them in bad wind sites), increasing them for some others that the government wanted to encourage, and granting additional bonuses for innovative technologies. Further revisions were made in 2008, 2010, and 2011, with tariffs for solar photovoltaics (PVs) and other technologies adapted to account for market development and cost reductions; most of these changes were part of an overall detailed monitoring and evaluation process that aimed to optimize the policy instruments.

Other policies to support renewables have included low-interest loans with favorable payment conditions, which have eased access to capital; changes to building codes to grant renewables the same legal status as other power generation technologies; a requirement that municipalities throughout Germany allocate potential development sites for wind power projects; continuing public R&D investment; training programs in various professions for renewable energy installations; and public information campaigns. In addition, in 2001 the German government introduced a tax on electricity consumption and raised taxes on fossil fuels (excluding coal); some of the revenues were used to support renewables.

Germany has seen rapid growth in renewable capacity and power generation as a result, with renewable energy's share of electricity generation rising from 3.1% in 1991 to 7.8% in 2002, 16.9% in 2009, and exceeding 20% in the first half of 2011. Wind power capacity has increased the most during this period, but solar PV and biomass power have also experienced substantial growth (see Figure 8.1) Biomass output increased particularly after 2002, when the FIT was extended to include cofiring, contributing more to annual output even without a capacity increase due to efficiency improvements. Biogas has also benefited from a FIT bonus for combined heat and power that favors the use of biogas, combined with favorable grid injection policies.

The focus of Germany's policies broadened in 2000 to encompass the heat and transportation sectors as well. The comprehensive "Market Incentive Programme" (Marktanreizprogramm, or MAP), was introduced to provide investment grants and soft loans for renewable heat systems. It provides grants to small systems and attractive loans for large systems. Over the years it was modified several times, but changes were discussed in advance with relevant industries, helping to increase acceptance of the program. It was also complemented with awareness raising campaigns and financial support for pilot projects to

Continued

CASE STUDY 12 Germany's Path toward Energy System Transformation—cont'd

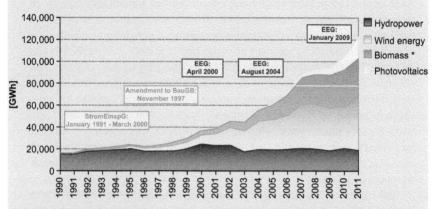

FIGURE 8.1 Development of power generation from renewable sources in Germany, 1990–2011. For color version of this figure, the reader is referred to the online version of this book.

promote R&D. It was supplemented in 2009 with a mandate requiring a minimum share of renewable heating and cooling in new buildings called the Act on the Promotion of Renewable Energies in the Heat Sector. To ensure that this would not set a cap on installations, the MAP is available only for projects that exceed the mandate. Although the MAP has driven growth in the heat sector, it is linked to the annual budget and has been temporarily suspended a number of times, leading to uncertainty in the marketplace and significant reductions in solar thermal installations in 2002, 2007, and 2010.

Biofuels were initially supported through tax exemptions, which stimulated significant investment. But they were gradually phased out and a biofuel mandate was introduced in 2007, leading to a sharp drop in consumption of biodiesel (especially pure vegetable oil). Biofuels are now supported by a tax exemption (to be phased out by 2014) and a mandate. In early 2011, the tax exemption for biofuels and quota fulfillment were made conditional on meeting sustainability criteria.

In the electricity sector, the FIT is widely believed to be the most important factor in driving growth in capacity and generation and reducing CO_2 emissions, and also to be responsible for creating an estimated 370,000 jobs as of 2010, advancing technologies, and helping to drive down costs. A report by Deutsche Bank Group (2011) found that Germany's schedule of automated price degression helped to drive down PV prices while also enhancing transparency and supporting security for investors. However, there has been increasing debate about the costs of the

CASE STUDY 12 Germany's Path toward Energy System Transformation—cont'd

FIT, particularly for solar PV. This was intensified by rapid declines in the cost of PV starting in 2009. FIT adjustments could not adequately correct for the rapid cost reductions. As a result, initially unscheduled downward adjustments were made, and they have been followed by price-corridor (or volume-responsive) degressions to ensure that tariffs adapt according to market development.

A decision in 2011 to phase out nuclear power by 2022 requires that Germany accelerate the transformation of its energy system. Based on the "German Energy Concept" released in autumn 2010, the government adopted, in reaction to the catastrophic events in Fukushima, a set of decisions for the transformation of the German energy system. The decisions maintained the ambitious but achievable targets for renewables set in the Energy Concept of 2010 (to meet at least 35% of electricity demand by 2020) and efficiency (20% reduction in primary energy consumption by 2020); in the longer term, Germany aims to have renewables account for 80% of electricity production and 60% of final energy consumption, and for primary energy consumption to be reduced by 50% (relative to 2008) by 2050. The decisions also include the acceleration of grid expansion and improvement of system integration, a roadmap and incentives for modernization of existing buildings, simplification of permitting processes for solar and wind power, exclusive economic zones for offshore wind, and coordination with states on elements such as development of criteria for onshore wind sites. The plan also highlights the importance of "the broadest possible support from society" and the need for all parts of society to work together for such large-scale changes to the nation's energy supply.

Key to development of renewables in Germany thus far has been the government's long-term commitment to renewable energy and the willingness to learn from experience. Ambitious near- and long-term targets—for individual end-use sectors and for the entire economy—have been supported by clear and consistent policies that aim to minimize risk for investors and have been flexible enough to adapt over time in response to market and technical changes. The FIT, in particular, has worked to minimize investor risk, encouraging significant investment from a variety of sources, enabling a broad range of people to participate in the process. As such, the FIT is credited with building strong and broad political support within civil society for both renewable energy and policies to support it. Early on, the linking of investment support with reporting on technology performance and clear monitoring mechanisms was helpful for learning about experiences in the field and advancing technologies. Efficiency targets and incentives have also played an important role in helping renewables to achieve a growing share of energy supply. Even in Germany, however, temporary suspensions (as with the MAP program) have slowed the market, highlighting the importance of maintaining consistent, long-term signals, as well as relying primarily on policies that are not tied to government budgets.

Continued

CASE STUDY 13 Denmark's Path toward a Sustainable Energy System

Prior to the first oil crisis in 1973, Denmark relied on oil for 90% of its primary energy supply. In response to the crisis, renewable energy and energy efficiency (in both energy production and use) became top political priorities, and the government has developed comprehensive energy plans on a fairly regular basis since 1976. Guided by these plans and long-term targets, a combination of policies addressing both the demand and supply sides has facilitated the development of renewable energy in Denmark.

Since the 1970s, a number of initiatives have focused on improving the efficiency of energy use across the economy, including the building sector, industry, and trade. Homes constructed in 2008 consumed only half as much energy per square meter as those built before 1977, for example. Efforts have also focused on expansion of district heat networks alongside the increased use of combined heat and power (CHP). As a result, the Danish economy has expanded by about 78% since 1980, while energy consumption has remained almost unchanged.

Over the years, Denmark has fairly consistently implemented policies to support renewable energy, particularly wind power and biomass for power and heat, and across all end-use sectors. Demand-side policies in support of wind power have included a gradually declining investment grant that was first enacted in 1979 and linked to required testing and certification procedures; a per-kilowatt hour subsidy for all wind power fed into the grid, enacted in 1985 and funded in part through a tax on CO_2; tax exemptions for wind power; agreements between the government and utilities to install specific amounts of capacity, including the world's first offshore wind farms, which helped utilities gain experience with renewables; and an FIT, which began as a voluntary agreement between utilities and wind producers and was fixed into law in 1992.

On the supply side, a publicly funded R&D program was launched in 1976, resource evaluations were undertaken, and the government later established a national wind power research facility. Interaction between small enterprises in the industry and the national test facility were critical for helping to improve knowledge about basic turbine design, thereby advancing technological development.

In addition to consistent R&D and market support policies that advanced technologies and reduced risk for investors, an important factor in wind power's success in Denmark has been broad participation in the market. Through the 1990s private investors, including small local cooperatives, owned more than 80% of the nation's total installed wind capacity. Widespread and local ownership helped to create broad-based support for renewable energy, and particularly wind power, because so many shared in its benefits. The government encouraged cooperative and local ownership directly through the use of incentives,

CASE STUDY 13 Denmark's Path toward a Sustainable Energy System—cont'd

such as special tax breaks, as well as ownership limitations (it is now possible to buy cooperative shares wherever one wants, and neighbors are compensated for visual- and noise-related impacts). Careful land-use planning also played an important role. In the early years, incentives for cooperative and individual ownership encouraged municipalities to set aside land for turbines; from 1992, national guidelines established regional capacity targets, designated specific areas offshore for wind turbines, and accelerated the permitting process, giving communities control over placement of projects while reducing uncertainty about project permitting and siting.

Denmark's progress with wind power halted abruptly in 2001 when the government shifted away from direct support for renewables and cut some R&D efforts, leading to a period of uncertainty; in addition, changes in planning structure delayed the siting of large wind turbines. As a result, little new wind power capacity was added over the next several years (see Figure 8.2). The wind market picked up again in the late 2000s, with the passage of new support policies (premium FIT for onshore wind, fixed-price tender for offshore wind, and fixed FIT for others) and particularly after the government announced a political target of eliminating fossil fuels from Denmark's energy system by 2050.

As of year-end 2009, the wind industry was Denmark's largest manufacturing industry, employing almost 25,000 people and accounting for one-third of

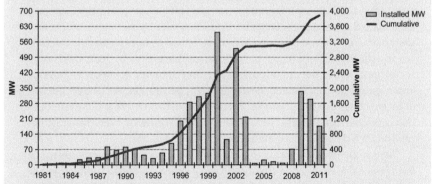

FIGURE 8.2 Installed wind capacity in Denmark, 1981–2009. *(From IPCC SRREN; To be updated or otherwise revised if kept.* - [2010 additions ~<300 MW for total of 3749 MW; 2011 additions estimated at 178 MW for total of 3871 MW, from GWEC, 2012.] For color version of this figure, the reader is referred to the online version of this book.

CASE STUDY 13 Denmark's Path toward a Sustainable Energy System—cont'd

the global market. Renewable energy accounted for about 19% of Denmark's total gross energy production, and renewables (especially biomass and wind power) provided approximately 28% of the nation's electricity. Denmark achieved all this while reducing its oil imports and national carbon dioxide emissions and increasing its energy exports. Since 1997, Denmark has been energy self-sufficient.

The story for biomass has been similar to wind power's, with a number of policies implemented over the years to shift from fossil fuels to a variety of biomass fuels (wood, waste, and straw) in the production of CHP. Three factors have been important for the development of biomass in Denmark, particularly agricultural biogas production and use. These include a bottom-up strategy that stimulated interaction and learning among key groups and the participation of a range of actors; consistent and continuous government financial support, including investment grants; and government support for decentralized CHP based on biomass and natural gas, combined with national energy taxes and the country's existing district heat systems.

Expansion of the district heat network has helped make possible this shift toward biomass-based heat and power. Over the years, there has also been ongoing expansion of the domestic electricity transmission network, as well as exchange connections with neighboring countries. This network has played an important role in enabling wind power to generate an increasing share of the nation's electricity. Other policies that have helped to promote renewables in various sectors of the economy include tax exemptions for renewable heat (biomass is exempt from the national CO_2 tax, and solar heat is exempt from CO_2 and energy taxes); CO_2 tax exemptions for transport biofuels since January 2005, and a mandate that a specific share of transport fuel sales be met with biofuels; and exemption from vehicle taxes and fuel consumption charges for electric vehicles.

To achieve the Danish goal of 100% renewable energy by 2050, the energy sector must be further transformed, not only through continued efficiency improvements and increased renewable energy generation (with the proportion of each dictated by costs), but also through new infrastructure and the introduction of flexible technologies (such as "smart grids" and electric vehicle infrastructure) and the design of integrated energy system solutions.

Keys to Denmark's success to date have included a public commitment to both renewable energy and energy efficiency improvements; a combination of policies (both "carrots" and "sticks") addressing the demand and supply sides; agreements with energy companies to deploy renewables in the beginning of the learning curve; R&D support (including publication of wind atlases) combined with technology guidelines and standards and learning from experiences in the field; development of necessary infrastructure; and policies to

CASE STUDY 13 Denmark's Path toward a Sustainable Energy System—cont'd

support local ownership and public participation in decisions about energy plans, technology development, and project siting. Also critical has been consistent, long-term and stable government support for renewable energy over most of the past four decades, particularly the FIT and other policies that have adapted to changing needs and circumstances, helping to minimize risk to a wide range of investors while increasing deployment, reducing costs, and creating broad-based support and a strong domestic industry.

CASE STUDY 14 Upper Austria on the Road to 100% Renewable Heat and Power

Upper Austria is one of nine Austrian states and home to about 17% of the nation's population, with 1.4 million inhabitants. Despite the fact that the region is highly industrialized—including machinery, metal production, wood processing, and automotive industries—in 2009, Upper Austria met more than 30% of its primary energy demand with renewable energy, including 45.6% of total heating demand and 78.4% of electricity demand. The region accounted for 25% of Austria's total installed solar collector area in 2009, is now the home to leading producers of renewable heating systems, and is on track to achieve its target (set in 2008) for 100% renewable space heating and electricity by 2030.

The government of Upper Austria put a priority on renewable energy and energy efficiency starting in the early 1990s (although a grant for solar thermal installations has been in place without interruption since 1981). The first energy strategy was passed in 1994 and led to an increase in the share of renewables from 25% in the initial year to 30% by 2000, with further energy and efficiency strategies enacted in 2000 and 2004, respectively.

The state's renewable energy and efficiency efforts have included a combination of policies to address both the supply and demand sides for technologies. Policies have been built on the three pillars of (1) "sticks" (regulations), including a renewable heat mandate, a low-energy building code, and simplification of permitting requirements; (2) "carrots" (fiscal incentives), including investment grants, which are tied to minimum quality standards and monitoring of output; and (3) "tambourines," or public awareness and training programs.

In addition to national support policies (including a FIT for renewable electricity) alongside Upper Austria's long-term perspective and clear targets with consistent, concerted, and coordinated policy measures, a number of elements have

Continued

CASE STUDY 14 Upper Austria on the Road to 100%
Renewable Heat and Power—cont'd

been key to the region's success. The government has kept an eye on the market to understand progress and then has adjusted policies as necessary. It began by targeting the most promising market sectors and expanded requirements after markets and technologies advanced; for example, in the heat sector it started with mandates for new construction, where installation of renewable energy technologies is easiest and least expensive, and then moved on to include buildings undergoing renovations and broader requirements. In addition, involvement of the energy agency in implementation has ensured that lessons learned from experience have been fed back into development of succeeding plans.

The region's public awareness and training programs have been critical for educating people about the benefits of renewable energy, for helping them understand their options and available financial instruments, and for creating a skilled workforce. With support from the O.O. Energiesparverband (State Agency for Energy Efficiency and Renewable Energy), inhabitants of the region have had access to energy consultations as well as vocational training programs through a regional Energy Academy. Training is provided to all actors along the value chain, from producers to installers of systems to users. The government established a solar research facility and a regional R&D program, and has led by example through government procurement programs of renewable energy systems. Publications, media and other campaigns, and competitions among municipalities have helped to raise public awareness and enthusiasm. In combination with regional policy support, such programs have helped to create strong local markets that have enabled companies to develop faster and to create more jobs, which in turn has increased support for renewable energy policies and increased confidence of regional politicians to continue their own backing of policies to support renewable energy.

Finally, renewable energy policies have been considered important as part of a broad policy agenda. Renewable heat, for example, has been linked closely to the building sector and to the need for energy-efficiency improvements. Further, such policies are considered to go beyond regional energy and environmental agendas to include agricultural (biomass heat and power), economic (job creation), and social policy, all of which has helped to create stronger alliances and broad public support.

In sum, success in Upper Austria is due greatly to a long-term government commitment to changing its energy system, with policy makers learning from experiences and feeding these lessons back into policies, public awareness and training programs that have helped to engage a broad base of society, linking of renewable energy to other sectors of the economy, and encouraging broad and deep support and interaction between the regional government and municipalities and the general public.

In all three, achievements thus far have resulted from a variety of factors, with some of the most important elements including long-term vision and public commitment to increasing the deployment of renewable energy across all sectors; a combination of policies to address both the supply and demand sides; policy frameworks comprised of both "carrots" (fiscal incentives and public financing) and "sticks" (regulations); consistent, stable, and predictable support, combined with flexibility and willingness to learn from experience and adjust policies and circumstances as required; and broad efforts to involve stakeholders, including advancing or encouraging public awareness and participation.

Moving forward, the challenges will include the introduction of more flexible energy system technologies to integrate even higher shares of variable renewable resources into the energy system; a broader cross-sector approach including many small and/or individual, distributed suppliers (households, industries, farms, etc.); storage (e.g., in electric vehicles or plug-in hybrids, pumped storage hydropower); interconnection across borders to share reserve and provide for trading when supply does not match demand; and conversion of the transportation sector.

A number of options can be used in combination to balance demand and supply. These include matching demand and the aggregated supply from a portfolio of renewable sources, can also include the following:

- A diverse portfolio of renewable energy resources and technologies, including dispatchable resources such as hydropower and biomass
- Geographic distribution of project sites
- Smart grid controls and intelligent load management, including flexible demand response
- Dispatch models that incorporate day-ahead forecasting for solar insolation and wind speeds
- A variety of very short-term and longer term energy storage options
- Fast-response natural gas turbines
- Grid interconnections with import to/ export from the system

It is important to note that in most cases it will not be practical or possible for countries to achieve full self-sufficiency, and trying to do so on a regional basis could increase costs of the renewable option. In the developing world, it will be critical to address issues related to affordability. At the same time, in rapidly growing developing countries, where there is no existing capital stock, there is the potential to build an energy-efficient infrastructure more quickly than in slow-growing developed countries.

8.1. POLICIES FOR TRANSFORMING THE ENERGY SYSTEM

A report prepared for the IEA-RETD, *OPTIMUM* (2011), examined the potential for improving system design to optimize use of renewable energy. It sets out a number of policies that are required for transforming the energy system, including the following:

- Adequate pricing on carbon, either through taxes or trade
- Accelerated replacement of capital stocks (including plants and machinery, vehicles, and power plants) through incentives or mandates to allow for investment in more efficient alternatives
- Energy-efficiency improvements through minimum standards, government incentives, and information campaigns
- More-sustainable urban and rural planning that encourages more-efficient buildings, greater density, walkable cities, public transportation, and so forth
- Public R&D to help push technological innovation
- Renewable energy specific policies to support market deployment, including a combination of "carrots" and "sticks" to support a diverse portfolio of technologies
- Planning and oversight to monitor progress
- Encouragement of consumer engagement and public acceptance through financial incentives or penalties, public awareness campaigns, demonstration programs, and so forth

Local Policies

Contents

Around the globe small hamlets and villages, cities, and urban communities—where the number of inhabitants ranges from a few thousand to several million—can develop their own strategies for renewable energy, and many have already done so.

Cities contribute to major social and environmental problems as they continue to grow; they currently account for about half of the world's population and most of the planet's greenhouse gas (GHG) emissions. At the same time, cities and other local communities have the potential to create substantial change, not only locally but also on the national or even global scale. This can occur by generating change through local transformations—through urban planning, building codes and design, choice of energy resources and technologies—in how they produce and use energy, and by providing lessons and creating innovative policies that can influence decisions in other towns or cities and at higher levels of governance.

Numerous cities and other local communities have adopted targets and enacted policies to advance the development and use of energy from renewable sources. They have been inspired by a range of drivers, including the desire for a secure energy supply, a cleaner local environment and healthier inhabitants, local sustainable development and job creation, and in some cases the need to broaden access to critical energy services.

Local governments face many of the same challenges and barriers that national and regional governments do. Examples of major obstacles include limited financial resources to pursue desired initiatives; a lack of financial, technical, and administrative support; and market pressures that ignore environmental and social costs and benefits in energy prices. But they have advantages as well, with their smaller populations, closer ties

Renewable Energy Action on Deployment
http://dx.doi.org/10.1016/B978-0-12-405519-3.00009-8

between governments and their citizenry, and generally minimal influence of powerful energy industries over policy makers. Perhaps most significantly, they have legislative and purchasing power that they can use to implement change in their own operations and in the wider community. As a result, it may be easier to introduce innovative and significant changes at the local level. Further, many of the opportunities for local policies exist because of the decentralized characteristics of most renewable energy sources, which enable energy to be produced at smaller scale and (in some cases) at the point of use. While centralized systems will continue to play an important role, renewable energy often finds its optimization in matching demand and supply close by.

This chapter discusses options available for local policy makers and highlights some "best cases." It provides recommendations for local policy makers, although policy makers at other levels of government can also profit from this knowledge. It is important to note that national and state/provincial policies can strengthen each other if designed to avoid inconsistencies and conflicts.

This chapter also draws from several publications and most particularly *the Global Status Report on Local Renewable Energy Policies* (May 2011) published by REN21 in collaboration with ICLEI–Local Governments for Sustainability and the Institute for Sustainable Energy Policies (ISEP).

9.1. OVERVIEW OF LOCAL OPTIONS

A wide range of local policies for renewable energy technologies has emerged in recent years, with a large variation in approaches. There are at least four categories of options available to local policy makers for advancing renewable energy production and use in their communities. These include target setting, local regulations and fiscal instruments, public procurement, and awareness raising.

Target Setting

Similarly to targets at the national level, local targets provide clarity and assurance for all stakeholders involved. While national targets can provide a starting point, local authorities can generally set targets that exceed national targets; in some instances, local targets are set before targets at higher levels of government. In most cases, these local targets cannot be binding for reasons of lack of jurisdiction, yet in practice the nonbinding character of targets does not affect the practical implementation at the local level.

The number of local communities with targets is large, and target options are numerous (see Table 9.1) For example, targets have been applied to operations of governments themselves, to inhabitants, to the whole city or province, or to specific subparts. They can mandate specific quantities or shares of energy to be derived from all renewable or from specific resources (such as the use of biofuels for road transport), they can require installation of renewable technologies (e.g., the installation of solar heaters in buildings through solar ordinances), or the use of renewable power by city or provincial offices. In many cases, renewable energy is seen as a critical tool for complying with broader goals for reduction of fossil energy demand or GHG emissions.

The most ambitious targets call for a shift away from fossil fuels to 100% renewable energy, generally by applying aggressive energy efficiency measures and supporting the deployment of renewable energy sources. In some cases, 100% targets cover only electricity or heat for local buildings, while elsewhere a 100% renewable target might include the transport sector (shifting to a combination of biofuels and renewable power) and industrial activities.

For small communities (i.e., <5000 inhabitants), targets to reach 100% climate neutrality might be achieved with the development of one large-scale renewable installation, such as a wind farm or a bio-energy installation that produces heat and power. A small but increasing number of communities are demonstrating that 100% (or nearly so) of energy demand can be met by a combination of energy efficiency improvements and renewable energy sources. Energy efficiency improvements assist in achieving such goals in a variety of ways, as discussed in Chapter 8; for example, they reduce energy demand and thus the associated costs of providing energy services with renewable energy. One such community is the region of Güssing in eastern Austria, which has created a new energy infrastructure, attracted numerous businesses and new jobs, drawn thousands of eco-tourists annually, and eliminated a large energy debt by becoming energy autonomous with local renewable sources (see Case Study 15).

Local Regulations and Fiscal Instruments

Local governments have access to a wide range of regulatory (such as building codes and urban planning) and fiscal instruments (e.g., local taxes) that can support a change in energy attitude. For example, a "solar ordinance"—an obligation to install solar thermal systems on new or renovated buildings—was first enacted by the government of Barcelona, Spain,

TABLE 9.1 Selected examples of local targets and incentives

Category/location	Target/policy
Share of renewable energy (all consumers)	
Boulder Colorado	30% of total energy by 2020
Cape Town, South Africa	10% of total energy by 2020
Stockholm, Sweden	80% of district heating from renewable sources
Tokyo, Japan	20% of total energy by 2020
Växjö, Sweden	100% of total energy (fossil fuel free) by 2030
Share of renewable electricity (all consumers)	
Adelaide, Australia	15% by 2014
Austin Texas	35% by 2020
Cape Town, South Africa	15% by 2020
Freiburg, Germany	10% by 2020
Sydney, Australia	25% by 2020
Taipei City, Taiwan	12% by 2020
Transport infrastructure and fuels	
Adelaide, Australia	Operate electric public buses charged with 100% solar electricity
Ann Arbor, Michigan	Subsidies for public-access biofuels stations
Betim, Brazil	Mandates for biofuels in public transport and taxis (plan through 2017); preference to flex-fuel vehicles for municipal vehicle fleet purchases
Calgary, Canada	B5 and B20 used in municipal fleet vehicles
Stockholm, Sweden	Plan to have 100% of all public transit buses run on biogas or ethanol by 2025; metro and commuter trains run on green electricity; additional biofuels stations
Building codes and permitting	
Barcelona, Spain	60% solar hot water in all new buildings and major renovations
Rajkot, India	Solar hot water in new residential buildings larger than 150 m² and in hospitals and other public buildings
Rio de Janeiro, Brazil	Solar hot water for 40% of heating energy in all public buildings

Continued

TABLE 9.1 Selected examples of local targets and incentives—cont'd

Category/location	Target/policy
Municipal electric utility	
Austin, Texas	Renewable portfolio standard 30% by 2020
Mexico City, Mexico	Net metering for solar PV
Oakville, Canada	Local utility voluntary green power sales

Source: REN21 Renewables Global Status Report 2012.

and has since spread to many other local and even national governments, especially in regions with high solar irradiation.

Other examples of these instruments exist, although they are not yet widely implemented. These examples include tax credits or abatement for residential installations, local biofuel mandates (e.g., for taxis in Betim, Brazil), and even a local feed-in tariff (FIT) for photovoltaic (PV) solar installations (e.g., Gainesville, Florida).

Public Procurement

Depending on local circumstances and specific characteristics of the local economy, public authorities have ultimate responsibility for a considerable share of total energy use. Not only do government facilities and equipment account for a significant portion of local or regional energy consumption, but local governments also generally have control over their own buildings, local public transport and other infrastructure, public lighting, parks, and harbors. In many cases they also control the local utilities for water, energy production, and sewage treatment. Thus, local policy makers and administrators have the potential to create numerous innovative combinations of efficiency improvements and renewable energy development.

Many cities are purchasing renewable power or biofuels for vehicle fleets or public transit vehicles. Others are incorporating renewable energy technologies into municipal infrastructure such as schools, hospitals, and other public facilities.

Public procurement helps to provide a stable and predictable market for renewable energy and, in some cases, to begin development of necessary related infrastructure and services (such as trained workers for installation, operation, and maintenance). In addition, it sets a strong example for others to follow, helping to educate businesses and the broader public about renewable energy options and thereby increasing local awareness and confidence in renewables.

In some cases, local governments go beyond their own public procurement and impose mandatory goals on the local infrastructure, managed by
municipal electricity utilities. For instance, the city of Austin, Texas, aims
to meet 35% of electricity demand with renewable energy by 2020, while
Mexico City has made it mandatory to provide net metering for solar PV.

Awareness Raising and Training

The "role model" function of the government, through public procurement,
is one avenue for raising the awareness of renewable energy opportunities
and technologies. In addition, policies will be most effective if the general
public and specific audiences, such as the local business community, understand the associated personal and community-based benefits that will result
from a shift to renewable energy, and if the resources, institutions, and skills
are available to bring this about.

Policies must be transparent and easily understandable to ensure that
target groups are aware of their existence and can act appropriately. Training and capacity building (e.g., with architects, electricians, city planners,
and other civil servants) can help to make developments more transparent
for specialists and people working in energy-related sectors. For the general public, information and promotion campaigns are most suitable. Public
education and focused energy audits are other options. A number of cities
have established information and demonstration centers to provide training and expertise on renewable energy and efficiency. They also induce
local innovation by bringing together relevant experts, small businesses, and
stakeholders.

Meanwhile, information must move in at least two directions: evidence
shows that public communication is not only about communication from
the government to its public, but also vice versa. The creation of strong public support for drastic changes in energy systems will require, among other
things, a government that listens to potential public concerns and that also
learns from local experience.

9.2. LESSONS LEARNED: FINDINGS FROM LOCAL POLICIES

It is widely acknowledged that local policies can play a key role in the
broader deployment of renewable energy. Some important findings can be
drawn from the experiences of ICLEI members, as set out in the REN21
local policies report. These lessons include renewables are important tools
for achieving sustainability, good ideas spread, size matters, framework

conditions are important, and communication is key. Each of these lessons is explained briefly in the sections below.

Renewables are Important Tools for Achieving Sustainability

In cities and communities with the greatest success to date, renewables (in combination with energy-efficiency improvements) are generally considered essential for achieving broad goals associated with economic and social sustainability over the medium and long term. Therefore, renewable energy often is a part in of broader policy packages for achieving economic and environmental sustainability.

Good Ideas Spread

Model cities are important, and their innovative policies and good ideas can catch fire and spread, both within countries and around the globe. For instance, the Barcelona Solar Ordinance, which mandates solar hot water in new construction, was copied by dozens of local governments in Spain and elsewhere (see Case Study 11).

In order to exchange information about innovative policy ideas and lessons learned, awards, competitions, and other instruments must be applied to increase national or international awareness of successful examples and best practices. Creating networks made up of local officials from a large number of communities can act as mentors and provide resources for others who wish to start similar activities in their own communities.

Size Matters

Small communities are highly flexible and their governments can enact policies and shift toward large shares of renewable energy more quickly than national governments can. They are generally more "outgoing" by actively seeking collaboration with other local communities. Cities of 100,000–500,000 inhabitants tend to be more active than smaller or larger communities.

Large communities—medium-sized cities and larger—are often reluctant to move toward 100% renewable energy targets. They tend to start by targeting specific opportunities, such as solar, wind, or bio-energy. This does not mean that larger communities cannot and should not aim for ambitious targets. While it may be more challenging at these larger levels—with larger areas of land to cover, more inhabitants, large industries, and other players to involve in the process—cities such as Rizhao, China (with nearly 2.8 million inhabitants), and regions such as Upper Austria (with more than 1.4 million inhabitants), are leading the way through a variety of measures combined with long-term government commitment (see Case Studies 14 and 16).

CASE STUDY 15   100% Renewables in Güssing, Austria

In the late 1980s, Güssing was faced with high bills for imported fuels and electricity that it had trouble paying. Located near the Hungarian border, this small agricultural town of about 4,000 inhabitants was one of Austria's poorest communities, and 70% of the workforce commuted weekly to Vienna due to a scarcity of local jobs. It hardly seemed a likely candidate to become a renewable energy pioneer.

By 2001, however, the town was energy self-sufficient, a magnet for eco-tourists, and an important industrial center due to the combination of substantial improvements in energy efficiency and deployment of renewable energy technologies. Producing biodiesel from local rapeseed and used cooking oil, and heat and power from the sun and through biomass steam gasification, the community reportedly became the first in the European Union (EU) to meet 100% of its energy demand with renewable sources.

Experts developed a revolutionary model in 1990 that called for a transition away from fossil fuels to the use of locally available renewables resources. The first step consisted of targeted measures to increase energy efficiency, starting with buildings in the town's center. Then the town began the step-by-step process of identifying local possibilities for generation of power, heat, and fuels; deploying renewable energy technologies; and organizing local stakeholders. By 1991, Güssing had its own biofuel production facility, and soon thereafter it added two wood burners for small-scale district heating systems. In 1996, with EU subsidies and financing from the province and the Austrian national government, the town built a wood-based district heating system to supply more than 85% of its inhabitants; an additional gasifier added within the next few years—for the production of gas, fuel, and heat—made Güssing energy autonomous.

The European Center for Renewable Energy was established in 1996. According to the municipality, an estimated 50–60 new companies were drawn to Güssing by the renewable energy supply, including Austria's first high-efficiency solar cell factory, all providing some 1,100 new direct and indirect jobs in the region and part of the profit was invested into renewable energy projects. Further, the district of Güssing, home to about 28,000 people, had plans to follow the town's path to energy self-sufficiency by the end of 2010.

The Güssing case shows how quickly things can change when a committed local government and other stakeholders—including energy experts, farmers, and a variety of businesses—come together in small communities to work for change. The town's step-by-step plan focused on energy-efficiency improvements and renewable energy deployment, combined with public financing from several sources, enabled the town to build a new energy system that is not only economically, socially, and environmentally sustainable, but has also brought higher living standards to the town's residents.

CASE STUDY 16   Rizhao: China's Solar City

Rizhao is a major Chinese seaport about 600 km north of Shanghai, and home to about 2.8 million people. This "City of Sunshine" is said to have received its name from a poem that refers to the "first that gets sunshine," and it is succeeding in living up to its name. Currently, Rizhao is well on its way to achieving its target of becoming 100% climate neutral.

In 1992, the city mandated that all new buildings install solar water heaters. While no subsidies were established for consumers, the local government provided the solar thermal industry with public funds for research and development to increase product efficiency and reduce costs. City officials and government offices led by example, being the first to install solar thermal systems. The combination of regulations and public education—through public television advertisements, assistance with panel installation, and open seminars, for example—led to increased interest in solar thermal systems and widespread adoption.

By 2007, almost all city households used solar water heaters, as did about 30% of all buildings in surrounding rural areas. Solar thermal panels heat greenhouses throughout the region, reducing energy costs for area farmers. Today, the tubular "Made in Rizhao" solar water heater reduces costs for the average household by about two-thirds during its lifetime, compared to the electric alternative, making it the best option for people in Rizhao. Moreover, the systems are manufactured locally, providing hundreds of jobs in the community.

The city has also supported energy-efficiency improvements while stimulating the use of biodigester technologies. Thousands of Rizhao area households also have solar cooking facilities, with many others using biogas digesters to produce cooking fuel from organic wastes. And solar PV systems power the city's street lights.

In recent years, the city's solar policies have resulted in new economic activities, while improved environmental quality has attracted an increasing number of tourists. In 2009, the United Nations recognized Rizhao as one of the world's most habitable cities.

Rizhao's success has resulted from a combination of strong political leadership, including leading by example; government support to advance development of a local industry and technology, and thereby improve efficiency and reduce costs; and the combination of use mandates and public education.

Framework Conditions are Important

National and state/provincial energy policies create the framework conditions to which local governments react. Such conditions could be targets for renewable energy, incentive programs, funds, FITs, competitions, or explicit funding for urban renewable development. Although it is not essential to have strong renewable support policies at higher levels of government for municipalities to succeed, the existence of such policies can facilitate developments at the local level.

Communication is Key

Creating "critical mass" and broad support is the way to bring renewables forward within the local community, both with professional stakeholders and with the public. Many model cities have established information and demonstration centers for renewable energy and energy efficiency to spread information to their own residents and beyond. Centers also provide training and bring together experts, small businesses, and stakeholders from their communities and further afield, to help spread knowledge and lessons learned to others.

Policies for Financing Renewables

Contents

Trillions of dollars worth of new private and public investments will be required to finance massive deployment of renewable energy technologies on a scale that will achieve a transition to a clean energy future within the next few decades. The necessary transformation is on the scale of the information technology revolution. However, the energy transition will be more challenging because energy infrastructure is embedded in a system that is highly capital intensive and highly networked. Policies will play a major role in tearing down the remaining barriers to renewables—related to maturing technologies, to a lock-in to existing infrastructure and to powerful vested interests in energy industries—and drawing new investments to renewable energy.

To date, investments in renewable energy manufacturing facilities, technology deployment, and infrastructure (see Chapter 1) pale in comparison to the capital needed over the coming years and decades (see Chapter 2). Given the current economic crisis in the countries in the Organization for Economic Cooperation and Development (OECD) and historic budget deficits in many countries, it is not likely that large public investments such as those necessary to spur the required shift will be available in the foreseeable future. Further, policies and other mechanisms are not currently in place to attract private investments at the scale required.

Justified or not, a potential investor still perceives the financing of renewable energy technology (or clean energy more broadly) as posing a greater risk, involving higher capital costs, longer time frames, and less certain rewards relative to many other potential investment opportunities. And yet, while public investment will remain critical, most of the new sources of finance must come from the private sector and existing financial markets.

To resolve this paradox a framework of strong, consistent, and long-term policies to support renewables and their associated infrastructure is required. In order to overcome challenges related to financing of renewable energy, it will be important to address a range of finance and nonfinance actions as part of a package of integrated solutions. In addition to using and improving upon existing successful mechanisms, new and innovative approaches and mechanisms will be needed.

This means that governments must view clean energy, including renewables (and energy efficiency), as part of a more comprehensive policy framework. Clean energy is not simply a part of the environmental strategy, but also a part of the economic development strategy, and one that could help lift the world out of the current economic crisis and provide new market strategies for even more aggressive growth. In the case of developing countries, renewable energy can be seen as a means to provide energy access, furthering economic growth while dramatically improving quality of life. No matter the reasons, such a shift in perspective has great implications for how renewable energy is financed and how public policy can be shaped to aid the transition to a clean energy economy.

This chapter is based on the recent International Energy Agency's Implementing Agreement on Renewable Energy Technology Deployment (IEA-RETD) study FINANCE-RE, which recommended steps as a path forward to finance large-scale deployment of renewable energy projects (September 2011).

10.1. BUILDING A NEW INFRASTRUCTURE

Transforming the energy system to a system largely dependent on renewable sources will require the development of new infrastructure. This includes the physical infrastructure needed made up of manufacturing facilities, generating units, connecting grids, and pipelines, and so forth; skilled workers and end users familiar with the technologies; and also the economic, legal, regulatory, and institutional infrastructure at the scale and scope necessary to attract financing for large-scale deployment.

In other words, radical infrastructure "transition management" is needed to bring about fundamental changes in sectors, if not entire economies. This implies new technical functions, new knowledge bases, and new organizational forms, and will demand a system- or economy-wide approach (see Chapter 8).

Further, breakthroughs in the cost, performance, and scalability of renewable energy technologies will be crucial. Government policy plays a critical role in making this happen, and governments and companies need to work together to craft policies and programs that will accelerate renewable energy innovation along the entire technology development value chain: from R&D to products and systems sold on the market. The development of solar photovoltaics (PV) shows how quickly costs can be reduced through government policies that drive a market, create competition and experience with production and installation, and attract private investment in R&D. (For an example of a program financing R&D efforts, see Case Studies 1 and 17.)

CASE STUDY 17 EU Framework Program/Marie Curie

The Marie Curie EU FP7 is the seventh framework in the EU for supporting collaborative research projects. International consortia and networks of excellence are provided financial assistance for developing new knowledge, technologies, products, demonstration activities, or common resources for research, with an emphasis on integration and exchange. Support is also provided for large-scale initiatives that require a combination of different funding types and from different sources (private and public, national and regional, or European).

It is important to recognize that existing infrastructure—ranging from roads, ports, and airports to electricity transmission and distribution systems—has come about as the result of creative, dedicated, and long-term public policy initiatives throughout the past century. The type and scale of effort needed, therefore, is not entirely unprecedented.

Potential New Sources of Investment

To achieve this scaled up level of investment, new sources of financing will need to be tapped. Beyond the traditional sources of renewable energy (and indeed energy more broadly) investments such as government programs, banks, development banks, and microcredit organizations, other potential sources include pension funds, sovereign funds, insurance funds, private corporations, and crowd funding mechanisms.

Pension Funds

Pension funds have been identified as potential significant players in the financing of renewable energy deployment and infrastructure on a large scale. A key target group is the "P8 Group," consisting of 12 of the world's largest public pension funds, with a combined USD 3.5 trillion in assets.

Historically, pension funds have not made infrastructure-based investments; instead fund managers have invested primarily in well-established funds and have been reluctant to take on project-specific risks. In some countries, such as the Netherlands, there are legal restrictions that limit the investment to stocks and bonds rather than to direct project investment.

However, pension fund managers are now searching beyond the traditional asset classes of equities, bonds, cash, and real estate. Pension funds are likely to play an important role in financing future renewable energy projects. Some may have appetites for strong growth and have the potential for investment in primary markets during the project construction phase; others with longer term steady returns could align with secondary markets and invest in the operation and maintenance of renewable energy projects.

There are some signs of increased interest, for example, by PGGM of the Netherlands and the Korean Teachers' Credit Union. Other institutional investors have committed USD 479 million to a China-focused infrastructure fund managed by the Macquarie Group and China Everbright. Macquarie is also raising USD 500 million for a North American renewable energy fund.

Sovereign Funds

The world's top ten sovereign funds have assets totaling approximately USD 3.8 trillion. The United Arab Emirates (UAE) holds the world's largest sovereign-wealth fund; its Abu Dhabi Investment Authority manages assets worth USD 627 billion. China has multiple sovereign funds totaling an estimated USD 831 billion; the largest is the SAFE Investment Company, with holdings worth USD 347 billion.

Many of the largest sovereign funds belong to oil exporters who are part of the incumbent infrastructure, although some of these governments (e.g., UAE, Saudi Arabia) are gaining interest in renewables. It may be easier to encourage sovereign-wealth funds in China, Singapore, and Hong Kong to finance clean energy infrastructure.

Insurance Funds

Given the correct financial or tax incentives, insurance companies could provide another new source of capital. The insurance and reinsurance industries are more concerned about climate change than are most potential renewable energy investors; some (such as Munich RE) are already investing in major renewable energy projects.

Private Non-energy Companies

An interesting new source of funding is emerging from some private non-energy companies that have started to take an interest in direct investment in renewable energy. For example, Google has invested more than USD 400 million in six renewable energy projects, either as conventional or tax and lease equity. The company's equity in the renewable energy generating plants allows it to sell power from these plants to the local markets as an "offset" for the electricity that its data centers consume in the same regions.

In 2011, Google decided to move away from its R&D program "RE<C" (or "renewable energy cheaper than coal"). However, Google continues to invest in deployment of renewable energy. In December 2011 the company announced plans to invest an additional USD 94 million in solar PV projects around California. This amount is on top of the USD 850 million that Google had previously invested in renewable energy deployment.

So far, Google is one of the first U.S. nonfinancial services companies to make use of tax equity incentives. But in the United States and elsewhere, tax incentives can be used to attract other profitable private corporations to invest in renewable energy and associated infrastructure, thereby contributing to the transformation of the energy system.

Crowdfunding

Although in many countries it is still marginal and dedicated specifically to funding of creative artists or musicians, the "crowdfunding" mechanism is gaining importance in entrepreneurship as well. The mechanism is based on attracting money from many individual small investors from the general public, which accumulates into a significant amount of money. In art and music, crowdfunding has been expressed as sponsorship, but crowdfunding investments in companies could be considered as loans, providing interest to investors if the company is successful. The use of crowdfunding has not yet reached a significant scale, and it is subject to investment regulations that differ from one country to the next. However, the mechanism shows great promise for smaller projects and companies it is promising, not in the least because crowdfunding enables members of the broader public to invest in renewable energy projects.

10.2. UNLEASHING NEW INVESTMENT: IMPROVING THE RISK-TO-REWARD RATIO

The key to unleashing the funds required, including potential "new" sources of funding such as those outlined above, is improving the (perceived) risk-to-reward ratio for investments in renewable energy. The risk-to-reward ratio compares expected returns of an investment to the amount of risk investors take on to capture these returns.

To this end, how can policy makers make renewable energy infrastructure more appealing to private investors? How can they make it perform like—or become even more attractive than—traditional infrastructure, such as industrial and municipal bonds? And how can investments in renewable energy deployment on a large scale generate competitive returns? And finally, how will this improved reward be revealed and assured to potential investors, removing any misconceptions about costs and returns?

Even in times of economic crisis, large amounts of money remain available for investment. Either they are invested in various forms of economic development—such as infrastructure (buildings, power plants, manufacturing equipments, pipelines, etc.)—or they are available in savings. For example, in the first quarter of 2010, U.S. corporations had more than USD 1.8 trillion in cash accumulated on their balance sheets; in the first quarter of 2011, foreign investment in Brazil reached USD 500 billion.

The challenge lies in attracting this money to the renewable energy sector. Investors have multiple options for where and how to invest their funds, and renewables must compete not only with other energy-related options but with the range of investment opportunities in all sectors of the economy. In many cases, costs of renewables are also overestimated. Compounding this challenge, many investors with deep pockets are currently in the "wait-and-see" mode, not confident about where or when to invest their money.

Government policies play a critical role in reducing real and perceived risks and improving the risk-to-reward ratio. One example of such policies is the "loan guarantee" (see Case Studies 19 and 21), with which the government guarantees, under certain conditions, to assume (part of) a debt that remains if a project developer, for example, is unable to repay a loan. Other examples of instruments that can help to attract private funds include tenders (see Chapter 5 and Case Studies 3 and 20).

CASE STUDY 19 U.S. Financial Institution Partnership Program Financing

The U.S. Financial Institution Partnership Program (FIPP) is a risk-sharing partnership between the Department of Energy (DOE) and qualified finance organizations that aims to expand credit capacity for renewable energy projects and to expedite the loan guarantee process. In FIPP financing, the U.S. DOE pays the credit subsidy costs of loan guarantees and provides a guarantee for up to 80% of a loan provided by qualified financial institutions.

Loan guarantees have been one of the dominant support mechanisms in the United States to facilitate private sector investment in deployment of large-scale renewable energy projects. The DOE has guaranteed more than USD 30 billion in loans and claims that these loans have saved or created more than 61,000 jobs.

Recent examples include guarantees to provide debt support to transmission infrastructure upgrade projects, and an offer of a conditional commitment for a USD 275 million loan guarantee to Calisolar Inc., at a former automotive plant, to commercialize its innovative solar silicon manufacturing process.

CASE STUDY 20 South Korea's Green New Deal

South Korea adopted its national policy vision "Low Carbon, Green Growth" in 2008, a "Green New Deal" designed to redirect about 2% of national gross domestic product (GDP) toward low-carbon initiatives. Most investment has focused on infrastructure projects, but South Korea announced in 2009 that it will invest USD 46 billion (or more than 1% of national GDP) over five years in clean technology sectors, with the explicit goal of increasing Korea's share of the global clean technology export market by 8%. The investment program will focus on solar PV, among other technologies (i.e., LED lighting, hybrid vehicles, and nuclear power).

The national vision consists of three parts: innovation, restructuring, and the value chain. Key elements of the program related to renewable energy include:
- Encourage business enterprises that specialize in overseas resources development.
- Encourage major strategic industries (such as automotive, chemicals, semiconductors, and steel) to increase green-tech R&D and capital investment. The objective is to increase green goods export in major industries from 10% in 2009 to 15% by 2013, and 22% in 2020.
- Re-think taxation, transportation, and energy infrastructure in order to capture the economic growth opportunity presented by renewable energy technology development.

Continued

CASE STUDY 20 South Korea's Green New Deal—cont'd

South Korea is also building local markets for renewable energy. For example, according to the government, 60% of individually owned houses (about 100,000 homes) were due to have solar panels by 2012, up from 14,500 in 2007. In addition, all government-planned housing facilities are to be equipped with solar PV panels.

CASE STUDY 21 The European Loan Guarantee Program

Under the umbrella of the European Investment Fund (EIF), the European High Growth and Innovative Small- to Medium-sized Enterprises Facility (GIF) supports entrepreneurship by investing (alongside with other private and public investors) in venture capital funds that focus on small, high-growth firms.

The EIF—which is owned by the European Investment Bank on behalf of the European Commission—invests in early-stage small- to medium-sized enterprises) and in funds that specialize in eco-innovation and have the option for proportionately higher European Union participation. The GIF is operated by the EIF and covers both early- and expansion-stage financing. Overall, it covers much of the life cycle of small, dynamic firms, and is also able to work with business angels, supporting their investments.

The GIF builds on the achievements of previous EU schemes, through which approximately USD 370 million (€309 million) was invested in 39 funds over the past decade, leading to investments in 357 small firms. The catalytic effect of these schemes has been substantial: EU public investment amounted to 17% of the combined total capital, attracting 83% of the required investment from private sources.

In addition to the use of specific individual policies, an integrated, economic-development approach to renewables will be required, with greater public and private sector cooperation and participation. This is a challenge at any time, and particularly in an era of economic crisis, when many governments face unprecedented public debt. And yet, continued and heightened public investment and policy support for renewable energy technologies can bring about new finance strategies and structures to stimulate economic growth and job creation, putting people to work in long-term jobs that will help build stronger and more sustainable economies for the future.

Some countries have already started down this path, making renewable energy a focus of sustainable economic development and international competitiveness. This is particularly true in some European countries and, most recently, in Asia, where an increased emphasis on clean energy development and policy support has emerged as a new form of "national industrial policy" in both China and South Korea. Increasingly, governments are recognizing that clean energy—and particularly energy from renewable resources—is not only essential for future environmental and economic security, but it can also provide enormous economic benefits.

The first movers among countries or even cities will likely benefit by becoming leaders in manufacturing, job creation, and further economic benefits. At the same time, many jobs and benefits are necessarily local, such as installation and maintenance. So even countries that are not early adopters will see major benefits with regard to local economic development and job creation, if they are able to improve the risk-to-reward ratio and attract more financing to renewable energy.

Drawing from Past Approaches

Existing infrastructure systems have relied on at least four types of targeted public and private approaches that can be employed again to improve the risk-to-reward ratio and thereby attract the needed funds to renewable energy. These include:

- **Economic development policies.** All investment that funded existing infrastructure resulted from policies that aimed to spur economic development and competitiveness. Such policies should address and connect the many actors throughout the economic system, while creating incentives and developing the case for a transition to clean energy.
- **Financial innovation and mechanisms.** Past public interventions to reduce risk for investors have made it possible for a diverse range of private investors to obtain safe, predictable returns, thereby drawing trillions of dollars in new capital for major infrastructure investments. Similar innovations can draw needed capital to the renewable energy sector.
- **Technology innovation policies.** Effective and efficient government policies have driven cost reductions and performance improvements in technologies and crucial enabling technologies that have then been integrated into the market at reduced costs. Employing such policies to advance renewables can perhaps ease the need for massive levels of future investment.
- **Stable public support policies.** Policies that create demand for existing and new technologies, mandate investment in infrastructure,

and support enabling environments provide stable, long-term signals that reduce risk for investors. These policies, which include fiscal
instruments, public finance policies, and regulations are discussed in
detail in the context of renewable energy in the preceding chapters
(see Chapter 5).

In turn, each of these strategies might require new institutional structures to promote and manage them. For instance, this will require policies
that internalize external costs; the pricing of carbon (by establishing a cap-
and-trade system or by carbon taxes), is one option. In addition, regional
and international cooperation on infrastructure investment and development or agreement on policy frameworks will facilitate concerted action
among different countries.

Pricing Carbon: Cap-and-Trade and Carbon Taxes

Europe's greenhouse gas (GHG) trading system (ETS) is the most advanced
and comprehensive scheme for pricing carbon and trading credits. But
while the ETS has emerged as a driver for GHG reduction via changes in
the fuel mix and some improvements in energy efficiency to date, the carbon market price has been too low to influence the level of financing for
renewable energy technologies.

Carbon taxes are an alternative to cap-and-trade systems. They enable
governments to specifically regulate the carbon price, while cap-and-trade
systems regulate a cap on emissions and leave the market to determine the
price. Carbon tax systems have not been broadly tested, but are expected to
be simpler and more flexible than cap-and-trade.

10.3. KYOTO MECHANISMS AS A SOURCE FOR EMERGING ECONOMIES

International mechanisms under the Kyoto Protocol regime have the
potential to attract investment in clean energy projects in developing and
emerging economies. However, these mechanisms have yet to prove their
promise for renewables. At current carbon prices, the Kyoto mechanisms—
the Clean Development Mechanism and Joint Implementation —have only
marginal impact on the financial attractiveness of most renewable energy
projects. Moreover, because these mechanisms do not provide payments
before project construction, they are not capable of directly tackling the
high up-front costs associated with renewable energy projects. High transaction costs exacerbate these challenges.

Longer term opportunities and threats are related mainly to future carbon market developments and to uncertainties regarding a future climate change regime (post-Kyoto Protocol), and will evolve in parallel with any new international climate regime.

Although some renewable energy technologies are already cost competitive with fossil fuels, and others are under some circumstances, higher carbon prices will help close the competitiveness gap. Carbon prices will ultimately depend on the type and level of emission reduction targets countries are willing to commit to, restrictions imposed on demand, potential other sources of cheap credit, the willingness of countries (or other actors) to voluntarily buy and cancel carbon credits instead of using them for offsetting, and the level of consolidation of the carbon market.

10.4. RECOMMENDATIONS FOR THE SHORT TERM: THE NEXT FIVE YEARS

In the short term, national governments should begin the process of adopting a broad package of support policies—including new policies related to economic development, finance, and technology innovation—in an integrated manner to support the build-out of the new clean energy infrastructure.

Specific recommendations under each category include:

Economic Development. Policies should support clear national economic development strategies that can attract step-change levels of capital to invest in a new clean energy economy. Infrastructure investment offers the economic development potential for nations to expand their economies.

- Identify gaps in various technology industry value chains and institute programs to fill those gaps, including manufacturing support, workforce development, and related industry support mechanisms.
- Create high-tech, clean energy clusters that optimize productivity by colocating different links of the supply chain (including R&D) and factors of production (supply of different components and a skilled workforce). Regional governments could administer the clusters and provide generous financial incentives such as grants, tax breaks, and discounted land to attract industry.
- Support business enterprises that specialize in overseas resources development to explore cooperative green-growth endeavors abroad.

- Develop regulations or incentives to convert existing industries (e.g., automotive, metals such as iron and steel, semiconductors) to low-carbon processes.
- Build aggressively local markets for domestic clean energy products.

Finance. Governments should aim to build a robust clean energy infrastructure that will attract more public and private capital investment for the near future and the long term.

- Institutionalize the functions to promote, integrate, coordinate, or manage the economic development, finance mechanisms, and technology innovation required for massive clean energy technology deployment, possibly under a new structure such as an investment bank, as in the UK (see Box 10.1 and Case Study 22).
- Create investment incentives with reassurances that will attract funds from a new and wider range of well-resourced investment pools—including tax incentives targeted to profitable corporations—and leverage public funds to achieve national goals.
- Explore the creation of "green bonds" to provide long-term, widespread capital for green infrastructure projects.

BOX 10.1 The Possible Advantages of a "Green Investment Bank"

- A green bank can align investment with mitigated risk and reduced risk-to-reward ratio; in doing so, it has the potential to target small numbers of larger scale projects to reduce transaction costs and drive down asset costs through savings associated with economies of scale.
- Respond to changing market needs—from the national level down to the level of individual, large-scale renewable energy projects—in real-time fashion
- Dilute investor risks by taking a first-loss debt position, or via guarantees or insurance-like products.
- Attract investments to the different risk profiles and to associated risk categories (construction and operational phases) for renewable energy projects.
- Enhance investment appeal using innovative financial tools such as up-front refinancing commitment, which guarantees an exit for long-term bank finance.
- Counterbalance investor fluctuations to stabilize market build-out with supplemental equity or debt.
- Include investment support in complementary support and services such as transmission, interconnection, and critical construction facilities.

CASE STUDY 22 The UK's Green Investment Bank

The UK government is preparing to launch its Green Investment Bank (GIB) in 2012, an innovative means for financing renewables and other green technologies; it will be the world's first such national bank. The GIB is expected to start as an "incubator" while awaiting European Commission approval to raise wholesale or retail debt funding. It will begin as a publicly owned bank for receiving investments, but it is expected to invest pooled public and private capital on a commercial basis and will ultimately be led by the private sector. The GIB is expected to initially bring in high levels of financial and technical expertise, working within the real market.

The UK government plans to fund the GIB with approximately USD 4.9 billion (€3 billion), and has suggested that the increased capital, much of which would be in the form of loan guarantees rather than subscribed equity, should enable the GIB to raise debt capital by issuing green bonds. This might, in turn, leverage a further USD 25 billion (€15 billion) in private sector funding.

The GIB proposes a suite of financial solutions, expecting to directly address risk mitigation by taking a first-loss debt position, or via guarantees or insurance-like products. Risk mitigation will be enhanced when combined with an innovative financial tool in the form of an up-front refinancing commitment, which guarantees an exit for long-term bank finance. Because the success of the investment in specific projects and the ability to attract additional investments depend on complementary goods and services, financial solutions will also involve suitable project infrastructure, transmission lines, interconnection, or installation vessels for offshore wind. Although it has yet to prove its value, the GIB offers the potential to reduce investor concerns about risk while simultaneously paving the way to attract new sources of finance.

Even before its formal launch, the GIB concept was being replicated. In May 2011, the Connecticut launched its Clean Energy Finance and Investment Authority (CEFIA), the first full-scale Green Bank at the subnational level in the United States. The Connecticut Clean Energy Fund is supported by a USD 0.1 cent per kilowatt-hour charge on electric bills. CEFIA will be responsible for developing programs to finance and support clean energy investment, manufacturing, research and development, and the financing of zero-and-low emissions power sources.

To a limited degree, some of the proposed GIB principles already exist in other investment institutions, including the European Investment Bank (EIB) and Germany's KfW. The EIB is committed to dedicating 20% of its financing portfolio to renewable energy projects, financing up to 50% of the investment costs of individual projects, a practice that significantly mitigates private investment risk. The EIB has played an important role in financing renewable energy deployment to date. In 2010, EIB's loans to the renewable energy sector accumulated to more than USD 8 billion (€ 6.2 billon).

- Align investment with mitigated risk and reduced risk-to-reward ratio that will, for example:
  - Target small numbers of larger scale projects to reduce transaction costs and drive down asset costs through savings from economies of scale
  - Respond to changing market needs—from a national level, down to the level of the individual, large-scale renewable energy deployment—in real-time fashion.
  - Dilute investor risks by taking a first-loss debt position, or via guarantees or insurance-like products
  - Attract investments to the different risk profiles and to associated risk categories (construction and operational phases) for renewable energy projects
  - Enhance investment appeal of renewable energy using innovative financial tools such as "up-front refinancing commitment," which guarantees an exit for long-term bank finance
  - Counterbalance investor fluctuations in order to stabilize market build-out with supplemental equity or debt
  - Include investment support in complementary support and services such as transmission, interconnection, and critical construction facilities

  Innovation. National governments should adopt policies to support technology innovation, which is necessary all along the technology development value chain, from laboratory to product, as well as in the commercialization stage.
- Develop more private and public R&D consortia (such as the EU's Marie Curie Program and the U.S. Sematech Program, Case Studies 17 and 18), along with a motivated venture capital market to support emerging technologies.
- Use "systems innovation" to increase innovation all along the technology development value chain, from lab to product development, to business and finance models; innovation is needed at all of these stages to increase performance and decrease costs of technologies.
- Use "open and distributed" innovation policies. These tap dispersed, global talent and enable people to collaborate across institutions through the use of internet tools and companies that link seekers and solvers on particular product development challenges. With these tools, researchers can supplement in-house efforts and accelerate the technology development cycle.
- Use "disruptive innovation" mechanisms to ensure that innovative technologies find success in early niche markets, where the fundamental

CASE STUDY 18 Sematech for Semiconductors

One nonenergy example that could be used in the renewable energy space is the Semiconductor Manufacturing Alliance (Sematech), initiated by the U.S. Department of Defense to strengthen the U.S. semiconductor industry. Sematech is global collaboration of semiconductor manufacturers who work together in the "precompetitive" research and development space to overcome common manufacturing challenges and reduce costs. It is one of the most unique and successful examples of industry and government collaboration.

This industry consortium, a not-for-profit membership organization, was started in the mid-1980s, when the U.S. government engaged 14 manufacturers to come together to solve common manufacturing problems and to collectively share risks associated with new industry processes. Funding for the research was shared by the private sector and the federal government.

While manufacturers were initially reluctant to work with their competitors, after 10 years the model was so successful that they continued to fund the collaborative research even after public funding ceased. The key was to identify the line between what is collaborative and what must remain competitive.

characteristics of the application are suited to the merits of the technology. For example, focus on potential new consumers who are looking for attributes that the current power system does not provide (such as high reliability or power quality).

- Look to "reverse innovation" strategies and partnerships, which entail designing, creating, and manufacturing climate technology products in developing countries to make them less expensive and then later adapt and export them to OECD countries.

Policies. Countries could consider a host of policies to support the scale-up of existing technologies and increase support for emerging technologies. For example:

- Adopt national or subnational feed-in tariffs or national tax credit schemes combined with mandatory renewable procurement for utilities, to achieve much greater penetration of renewable power into the existing domestic generation mixes.
- Encourage more turnover and avoid technology lock-in. For example, adopt or require use of a leasing model, in which a developer retains project ownership (generally of distributed technologies like solar PV) and leases the technology or sells resulting electricity under a long-term contract with rates similar to the retail price of electricity.

- Create subnational, special-purpose funds to support renewable energy and energy-efficiency investments; these funds could be financed through a modest surcharge on utility bills, although they may have a more specific source of funding (e.g., a negotiated settlement with a utility).
- Mandate (increased) public procurement of renewable power from government agencies. Defense agencies, for example, are often the largest energy consumers in a country and have enormous procurement power.
- Implement mandatory use of renewable technologies in new buildings, such as solar electric or solar water heating or other renewable technologies; this can be accomplished by amending building codes.
- Establish an "emerging technology renewable auction mechanism" (ET-RAM) that would require locally regulated utilities to procure clean energy project outputs from specific technology classes up to a predetermined cost limit, at guaranteed prices competitively bid by the winning developers, in order to address "Valley of Death" commercialization gaps. Design mechanism to overcome concerns about available demand and price levels that typically face efforts to finance emerging technologies.
- Along with the ET-RAM, either mandate or encourage insurance companies, through fiscal incentives, to provide "efficacy insurance" to provide protection against a technology that does not perform as projected. Some form of backstop "reinsurance pool" perhaps guaranteed by the national or subnational governments, could complement this product.

Outlook to the Medium and Long Term (2020–2050)

Once new programs are established, the focus turns to implementation. When proof of rewards is demonstrated by early large-scale projects, and initial returns are reinvested for subsequent projects, more cautious investors will become engaged and levels of funding will increase with each cycle. In the longer term, national goals can start to increase if they are aligned with technologies, enhanced investment levels (driven by mandatory levels through quota policies or Renewable Portfolio Standards or procurements with minimum public financial support), and attractive investment conditions conducive to a newly established clean energy infrastructure.

Historically, governments have funded clean energy projects primarily to advance climate and environmental objectives, with economic development viewed as a secondary benefit. The path forward needs to place economic development at the forefront of creating a renewable energy infrastructure for the twenty-first century. Countries must view renewable energy as a challenge of financing infrastructure for the near and long terms, and need to align the risk-to-reward investment profiles in a way that begins to resemble conventional infrastructure finance. A strategy that combines targeted and comprehensive programs in the four areas of economic development, finance, innovation, and public policy is required to create a strong foundation for the transition to a sustainable energy economy.

CHAPTER ELEVEN

Getting on Track: Lessons Learned for the Road Ahead

Contents

11.1. THE URGENCY FOR ACTING NOW!

The need for changing our energy systems is clear. With volatile oil prices that are trending upward and unstable sociopolitical situations in several key oil-producing countries, even as energy demand continues to rise around the world, the security of energy supply and hence the global economy is at risk. With ever increasing greenhouse gas (GHG) emissions, even during a period of economic crisis in Organization for Economic Cooperation and Development (OECD) countries, stability of the global climate is at risk. The Deepwater Horizon oil spill in the Gulf of Mexico in 2010 and Japan's Fukushima nuclear crisis in 2011 have made it even more evident that the other security, environment, and human health risks of our conventional energy system are very real, devastating, and extremely costly.

Scientists caution that global GHG emissions must peak within this decade. Further, they must drop by 80–95% by 2050 (relative to 1990 levels) in order to limit warming below the internationally agreed 2°C threshold and thereby avoid the worst impacts of climate change. At the same time, it is critical to avoid a lock-in to energy technologies that not only contribute to climate change but that also offer no long-term energy security.

There is little time for this transition, as the world's increasing demand for energy requires substantial investments in new production capacities and infrastructures. Each year that passes, and each dollar invested in existing or new fossil energy technologies and infrastructure, is further locking the world in to dominating "old" energy systems and to catastrophic climate change.

Renewable Energy Action on Deployment
http://dx.doi.org/10.1016/B978-0-12-405519-3.00011-6

Renewable energy technologies and energy efficiency improvements will both play major roles in shaping our energy future. Both can make important contributions individually, but the special synergies of renewables and energy efficiency will make it easier and cheaper to substantially increase shares of renewables in the energy mix by working in combination. They can also be important drivers for economic development. Many renewable technologies are already experiencing dramatic rates of growth, particularly wind and solar power. But most are growing from a relatively low base and, at current rates of deployment, society may not be able to reap the full economic, climate, and other benefits and opportunities that these technologies offer.

Despite the current economic and fiscal crisis, and limited budgets available for driving the necessary change—or perhaps because of them—decision makers inside and outside of government, should take the responsibility today for fundamentally improving the energy system of tomorrow. The economic benefits of such a strategy can emerge faster than most people think.

11.2. THE ACTION STAR FOR DECISION MAKERS

The International Energy Agency's Implementing Agreement on Renewable Energy Technology Deployment (IEA-RETD) recommends six practical and realistic categories of policy options for the next five years, graphically represented by the six-pointed ACTION Star. Ideally, all six actions will be included in the policy portfolio in order to facilitate a quick start by removing major barriers to deployment and attracting the required financing to renewable energy and associated infrastructure. Policy makers can choose to give priority to one or more particular actions as needed.

IEA-RETD and other studies analyzing experiences with policies for renewable energy reveal four overarching guidelines for getting on track:

- Combining renewable energy deployment with improvements in energy efficiency is essential to a cost-efficient transition.
- Maximizing benefits and thus support for renewable energy are key. Broad stakeholder participation in planning or ownership of projects can minimize opposition (reducing risks and costs), increase available funds for investment, and increase public and political support for renewable energy.
- Designing a robust financing framework that can withstand economic crises is critical for improving the attractiveness of investing in renewable energy and associated infrastructures within changing landscapes.

- Overcoming the inertia and acceptability issues will be a "game changer" as there are powerful forces working against a transition toward an energy supply based predominantly on renewable energy.

Policy makers can yield near-term results by applying six key lessons. These actions are not necessarily in order of priority. Ideally, a policy portfolio contains all six elements:

- **Alliance building** to lead the paradigm change
- **Communicating** and creating awareness on all levels
- **Target setting** at all levels of government
- **Integrating** renewables into institutional, economic, social, and technical decision-making processes, while integrating renewable policies with efficiency policies.
- **Optimizing** and applying proven policy instruments
- **Neutralizing** disadvantages and misconceptions

⭐ Alliance Building to Lead the Paradigm Change

The transition to a sustainable energy system based primarily on renewable energy cannot be achieved solely from the top down. A collaborative effort is needed to overcome powerful forces working to maintain the status quo, as well as lock-in to existing infrastructure, technologies, and mindsets, all of which create inertia and slow the pace of change. Collaborative efforts among policy makers and stakeholders will also help to ensure that policies are designed and implemented effectively. Thus, new alliances are required, among policy makers, investors, environmental organizations, businesses, communities, households, countries, and regions.

The transition toward a truly sustainable energy system will have a major impact on human society. It will bring significant and broad-reaching benefits in the short term, and particularly in the long term, compared with the costs of continuing on the current energy path. Thus, it is imperative that action begins NOW. At the same time, the transition will not be an easy process. Political decisions are generally taken based on short-term needs, especially in times of economic crisis; there are powerful forces working to maintain the status quo; and lock-in to existing infrastructure, technologies, and mindsets create great inertia that slows the pace of change.

Large-scale deployment of renewable energy will demand a conscious effort by all stakeholders involved, and a large increase in the number and breadth of stakeholders—policy makers, businesses of all sorts and sizes, communities, and individuals. Given the global challenges faced, a collaborative effort is required. Hence, alliances need to be built among countries and within regions; between governments and the private sector, including energy companies and utilities; between businesses and non-governmental organizations (NGOs), and communities, and so on. Such alliances can underline the crucial role of renewable energy and energy efficiency in future energy systems, and can help to shape this future.

How can governments and businesses create effective alliances? What examples of successful models can be followed? Are there successful examples of engaging utilities and traditional energy companies in the renewable energy sector and changing mindset such that they see renewable energy as mainstream?

Examples of policies and strategies:

- Green alliances are already being formed, either as green industrial alliances, public–private cooperation in promotion of and communication on renewable energy, "green deals" between industry and government, cooperation between environmental NGOs and industry, or in other forms. The role and commitment of governments can be increased, for example, by reaching out beyond the most obvious stakeholder groups (such as the main energy producers, and primary industrial and service consumers).
- Assess the levels of support for renewable energy technologies among different stakeholder groups, and the drivers for opposing or supporting them.
- Find creative ways, through existing policies or innovative new ones, to address stakeholder concerns and to encourage broad (including local) ownership and investment strategies for renewable energy projects and related infrastructure.

Communicating and Creating Awareness on All Levels

It is important for decision makers across societies to recognize the potential, opportunities, and benefits of renewable energy and to have the knowledge, workforce, and skills to realize them. This requires continuous consultation among stakeholders about their experiences and interests, as well as multidirectional communication about issues and benefits of renewable energy. The broader public (including policy makers) needs to understand the full economic and social costs of the current energy system (including external costs) and the renewables potential to provide a growing share of energy supply. Misconceptions need to be corrected with accurate information. Policy makers on

all levels need to understand what is required to attract investment and to advance renewables. Policies and regulations need to be widely known, transparent, and easily accessible to relevant actors. Renewable resource potential needs to be measured and information made accessible. Further, a skilled workforce is required.

Examples of policies/strategies:

- Partner with the private sector to gain understanding of renewable energy technologies (and related infrastructure needs) and challenges to the industry in order to develop relevant or innovative support policies.
- Work with media (radio, television, newspapers) on public service announcements regarding potential of renewables, existing policies, and policy changes (where relevant), and make information easily accessible through the internet.
- Use the building permitting process to target relevant decision makers and educate them about renewable energy policies and systems.
- Establish training programs for skilled workers across the supply chain, and also in related fields (architects, plumbers, city planners, etc.) at regional and local levels.

Target Setting at All Levels of Government

Ambitious and realistic long-term and interim targets, preferably binding, at different levels of government provide the desired predictability to the energy market. Such targets need to be grounded on clear general goals, and need to be advanced by specific renewable support policies that promote not only the deployment of renewable technologies but also the development of needed infrastructures. This will create a stable and favorable investment climate for renewable energy technologies. Countries like Germany, Denmark, and China are proving the effectiveness of ambitious targets in combination with strong support policies, as are many local governments.

Deployment can be ensured by targets at the local, national, and international levels, as well as targets by sector (i.e., power, transportation, heating, and cooling).

Examples of policies/strategies:

- Explore the opportunities for target setting among groups of countries at similar levels of economic development, such as the G20 or OECD framework.

- Explore the opportunities for regional target setting, for example, for South America or North Africa.
- Encourage businesses to set internal targets or partner with others to establish broader targets and share purchasing, and so forth.

Integrating Renewables into Institutional, Economic, Social, and Technical Decision-Making Processes

Since the 1980s, many countries have worked to integrate the environment into various areas of policy making and infrastructures; the same should be done with renewable energy, taking advantage of synergies with energy efficiency. By integrating renewables into building and construction codes or policies to support, for example, innovation, finance, urban planning, and broader economic development, renewable energy will achieve a higher status that will, in turn, help to reduce regulatory inconsistencies and barriers to their deployment. The same applies to the technical integration of renewable energy technologies into grid systems and other infrastructure.

Renewable energy manufacture and deployment have the potential to create new jobs and economic growth, providing a strong business case for their acceleration. However, in order to bring this about, governments face the challenge of developing a consistent set of policies along every segment of the technology value chain: from research to technology innovation, to manufacturing, to domestic deployment, to infrastructures and to export markets.

One of the most significant conclusions of RETD work is that renewable energy should be considered part of the whole institutional, economic, and infrastructure system, instead of a separate area of policy. By incorporating renewables into other or broader areas of policy, regulatory inconsistencies and barriers to renewables can be reduced, enabling progress to occur in a more efficient, effective, and sustainable manner. Broad public participation is also important to ensure that stakeholders at all levels have a voice in the process, thereby minimizing opposition and maximizing support for renewable energy.

Such an integrative approach is important for attracting the massive financial flows and the strong, broad public support that will be required for a significant scale-up in renewable energy manufacture and deployment.

While integrating renewables into systems and infrastructure, quality standards and certification criteria are also required in order to prevent low-quality technologies, installations, or unsustainable production (e.g., of biofuels), and to develop consumer confidence.

Examples of policies and strategies:

- Improve the organization and regulation of energy markets and infrastructures in order to enable system optimizations and an increased uptake of renewable energy (e.g., related to smart grids, the emergence of "prosumers" (consumers that are also becoming producers). The more mature and cost-effective renewable energy technologies should be treated as mainstream technologies, rather than as niche technologies that are marginal in the energy system.
- Enact regulations that require renewable energy technologies (and energy efficiency improvements) as a component in new and existing buildings.
- Use the model of energy service companies for the integration of renewable energy in the renovation of public buildings.

Optimizing and Applying Proven Policy Instruments

It is important to build on one's own positive experiences and to learn from the experiences of others. Proven policies can and should be adapted and optimized to national or local circumstances and needs. In those countries with existing renewable energy support policies, IEA-RETD studies provide evidence that continuing on a consistent policy path—for example, improving an existing policy to address specific concerns and to meet changing circumstances—is generally more effective than switching policies midstream. This is because investors gain confidence from stable and predictable policies.

Countries that have continued to use and improve upon specific policy mechanisms for an extended period of time (at least beyond the relevant horizon of businesses) have attracted substantial private investment in the manufacture and deployment of renewable energy technologies. A continuation of these successful strategies may result in the best conditions for a continued and increasing deployment: stability. However, this does not exclude the possibility that in some cases it would be better to introduce new policies.

But how could the policy design respond to the new economic and financial context? What are the most effective and efficient policies, in terms of (government) budgets and market development? How could policy design contribute to getting out of the economic crisis? How could these policies evolve if renewable energy is more and more determining the characteristics of the energy markets? And which lessons have to be learned for optimizing the effectiveness and the (cost) efficiency of policies?

Examples of policies and strategies:

- De-risk investments in renewable energy by introducing risk-removing or risk-sharing financial instruments (such as government-backed loan guarantees and insurance), and through public participation in projects.
- Keep financing of renewable energy support schemes outside of the government budget.
- Use power of public procurement. In addition to helping create stable demand, public procurement enables governments to lead by example, increase demand, and help drive down costs through economies of scale, and to create greater public awareness and acceptance.

Neutralizing Disadvantages and Misconceptions

Energy markets are built on numerous government rules and (financial) interventions, many of which are not consistent with the goal of a low-carbon energy system; further, renewables are trying to compete on an uneven playing field. Moreover, misconceptions exist about costs and risks of renewable energy, among other issues. Integrating external costs (and benefits)—such as climate change and health impacts—into energy prices will help to level the playing field for renewables, helping to improve the economic competitiveness of renewable energy and significantly increasing investment in this sector. This step can also reveal the real costs and benefits associated with increasing the deployment of renewable energy.

The "true" societal costs of our current energy system are not fully reflected in the prices that consumers pay. First, costs related to environmental (e.g., climate change) and social impacts of energy production and use, as well as the security of energy supply, are not properly reflected in current energy prices. Although it is difficult to determine what these actual costs are, they are not zero, and various policy instruments (such as taxes on pollution or on natural resource use) exist to address this market imperfection.

Second, enormous subsidies remain for fossil fuels and nuclear power, despite the fact that these are mature fuels and technologies and come with very large "external" costs. It is not consistent to incentivize the deployment of renewable energy technologies and, at the same time, continue to support (mature) conventional energy technologies, thus preventing a level playing field.

Examples of policies/strategies:

- Develop an inventory of existing support mechanisms for fossil fuels, such as direct subsidies or tax reductions, to identify existing distortions in the playing field.

- Make transparent the amount of government support (direct and indirect) that is provided to the energy sector (by technology, end-use sector, etc). This will strengthen the case for strong government policies to support energy efficiency and renewable energy technologies.

- Level the playing field for renewable energy by removing subsidies for fossil fuels, by setting a gradually increasing price on pollutants (GHGs and others), and by introducing or modifying fiscal instruments to internalize other costs and benefits associated with energy production and use, such as pricing carbon, by taxing CO_2 or establishing an emissions trading system.

Players in the Field

An overview of the most relevant stakeholders in the global field of renewable energy technologies deployment.

The International Energy Agency (IEA) was established in 1974 in the wake of the first major oil crisis. The initial focus was on oil supply, but it has since broadened to include renewable energy, among other issues, with an emphasis on gathering of statistics, analysis of policies and markets, and addressing issues related to systems integration. The IEA acts as policy advisor to its member states (OECD countries), but has also worked with non-member countries (particularly China, India, and Russia) through implementing agreements. The IEA-RETD, the initiator of this report and one of the IEA's implementing agreements, seeks to maximize the contribution that renewables can make to mitigating climate change, improving security of energy supply, and advancing economic growth. The IEA's International Low-Carbon Energy Technology Platform was established in late 2010 at the request of G8 and IEA ministers to encourage and accelerate the deployment of low-carbon energy technologies, including renewables.

The International Renewable Energy Agency (IRENA) was founded in early 2009 and, in April 2011, the inaugural session of the Assembly established IRENA's first work program with an initial budget of USD 25 million. IRENA's mission is to promote the adaptation and sustainable use of all forms of renewable energy. Member States have pledged to advance renewables through national policies and to advance the transition to sustainable and secure energy supply through domestic programs and international cooperation. Central to its mission is IRENA's objective to become a global clearing house for renewable energy knowledge by facilitating access to technical, economic, and resource data, while also sharing and building on existing knowledge of best practices and lessons learned on renewable energy policy, capacity building, finance, and relevant energy-efficiency measures.

The Renewable Energy Policy Network for the 21st Century (REN21) was established in 2005 following the International Conference on Renewable Energies in Bonn (2004). REN21 connects governments, international institutions, industry associations, nongovernmental organizations (NGOs), and other renewable energy partnerships and initiatives with the aim of enabling a rapid transition to renewable energy. Its annual Renewables Global Status Report provides a unique overview of the worldwide status of renewable energy markets, industries, investment, policies, and rural energy developments.

The United Nations Framework Convention on Climate Change (UNFCCC) was adopted in May 1992. Its aim is to reduce greenhouse gas emissions to avoid dangerous global climate change, and to cope with whatever temperature increases are inevitable, through an intergovernmental framework that is enforceable at the national level. The Kyoto Protocol is an international agreement linked to the UNFCCC, and its various mechanisms—including carbon trading, the Clean Development Mechanism, and Joint Implementation—rely on renewable energy, energy efficiency, and other technologies and options to reduce emissions.

The Renewable Energy and Energy Efficiency Partnership (REEEP) aims to facilitate the transformation to more secure and sustainable energy systems that reduce greenhouse gas emissions and improve access to energy, and thereby wealth creation, for the world's poor, by accelerating deployment of energy efficiency and renewable energy technologies.

International Renewable Energy Conferences (IREC) are a series of ministerial level conferences on renewable energy that began in Bonn, Germany, in 2004. Results of the first conference included a Political Declaration with shared political goals for an increased deployment of renewable energy, an International Action Program of voluntary commitments by governments and other stakeholders to specific renewable energy targets, and Policy Recommendations for Renewable Energies to guide policy makers. Subsequent meetings have been held in Beijing (2005), Washington, D.C. (2008), and New Delhi (2010). Additional pledges were made by national and local governments in New Delhi.

The World Energy Council (WEC), founded in 1923, is a global forum that covers all forms of energy and aims to promote the sustainable supply and use of energy for the greatest benefit of all people. It has member committees in 91 countries. Objectives include gathering and disseminating data, and promoting research about energy with the greatest social benefit and lowest possible environmental harm; convening meetings and workshops to facilitate the supply and use of such sustainable energy; and collaborating with other like-minded organizations.

G8/G20: The international meetings of the largest economies in the world are growing increasingly committed to mitigating climate change and to fostering clean energy, green growth, and sustainable development. To advance these goals, the G20 Clean Energy and Energy Efficiency (C3E) Working Group were created; together with the IEA, this working group reported about related developments at the 2011 Cannes G20 meeting.

The Roundtable on Sustainable Biofuels is a multi-stakeholder initiative of the Swiss École Polytechnique Fédérale de Lausanne that aims to develop international standards for the sustainability of biofuels and to create and implement a certification system based on these standards.

The Global Bioenergy Partnership (GBEP) is a partnership that brings together stakeholders from governments, the private sector, and civil society to promote bioenergy for sustainable development, with a focus on developing countries. It focuses on three strategic areas: sustainability, climate change, and food and energy security. Its purpose is to help organize and implement international R&D and commercial activities. The GBEP also supports the development of policy frameworks.

A.1 INTERNATIONAL FINANCIAL INSTITUTIONS AND NATIONAL DEVELOPMENT AGENCIES

World Bank Group (WBG): This group consists of the International Bank for Reconstruction and Development and the International Development Agency (which together make up the World Bank), as well as the International Finance Corporation (IFC), the International Center for Settlement of Investment Disputes, and the Multilateral Investment Guarantee Agency. Together, they support efforts by developing countries to provide reliable and cleaner electricity to businesses and households through financing instruments, knowledge transfer, partnerships, and policy advice. In 2004, at the Bonn International Conference on Renewable Energies, the WBG committed to increasing its financial support for new renewable energy (excluding hydropower >10 MW) and energy efficiency at a growth rate of 20% annually between 2005 and 2009. In actuality, the WBG far exceeded that commitment (with more than USD 7 billion over the period rather than 1.87 billion). In 2008, under a Strategic Framework on Development and Climate Change, the World Bank Group committed to increasing investment by a further 30% annually during fiscal years 2009–2011, reflecting increasing priority to renewables and energy efficiency for alleviating poverty and achieving sustainable development.

Regional and national development banks: In addition to the WBG, other international development banks are financing renewable energy projects to promote economic and social development in their specific regions of the world. These include the Asian Development Bank, the Inter-American Development Bank, the European Bank for Reconstruction and Development, and the African Development Bank. They each provide long-term loans at market rates, very long-term loans at below market rates, and financing through grants. There are also a number of subregional multilateral development banks, as well as national development banks such as Brazil's BNDES, Germany's KfW, and the China Development Bank. According to Bloomberg New Energy Finance, 13 development banks around the world provided USD 13.5 billion in financing for renewable energy projects during 2010, up from USD 8.9 billion in 2009, and triple the USD 4.5 billion in 2007. The top three were the European Investment Bank, the BNDES of Brazil, and the KfW, each of which provided more direct project finance to renewable energy than did the World Bank Group in 2010.

Global Environment Facility (GEF): Established in 1991 as a pilot program in the World Bank, the GEF was restructured and moved out of the World Bank system in 1994. The GEF partners with international institutions, NGOs, and the private sector to address global environmental issues, including climate change, biodiversity loss, and land degradation. Its partners include the United Nations Environment Program, UN Development Program, the World Bank Group, and regional development banks, among others. Renewable energy projects supported by the GEF aim to bring about a step-change in the development and diffusion of least-cost technologies. In 2010, funding was approved for 25 projects with a direct GEF contribution of USD 40.4 million and total cofinancing from all sources of USD 382 million.

A.2 NON-GOVERNMENTAL ORGANIZATIONS

World Business Council for Sustainable Development (WBCSD): The WBCSD is a global association of businesses whose objective is to promote the business case for sustainable development and to participate in policy development. Recently, the Council has called for clear signals toward a low-carbon economy and a level playing field across international markets, which could best be achieved through global and stable carbon prices.

Greenpeace: The international environmental organization Greenpeace has been working for well over a decade to advance renewable energy, coproducing blueprints with other organizations (such as Wind Force 10 and Solar Generation). Perhaps the best known of its renewable energy reports is the Energy [R]evolution series, produced in collaboration with the German Aerospace Center (DLR) and other contributors, which has outlined aggressive energy scenarios to reduce CO_2 emissions by at least 80% by 2050 and provides guidance for policy makers.

World Wide Fund for Nature (WWF): This organization began to focus on fossil fuels and climate change in 1990s, and climate change became a priority in the 2000s. In early 2011, in collaboration with Ecofys and the Netherlands-based Office for Metropolitan Architecture, WWF released a report that examines the feasibility of achieving 100% clean and renewable energy by 2050.

A.3 PRIVATE SECTOR—INDUSTRY GROUPS AND OTHER PLAYERS

Renewable energy industry associations: In the past several years, the number of international industry associations has increased significantly, while some regional groups have expanded their focus to global issues. Such groups generally focus on some or all of the following: data collection, research, education or policy advocacy, and the provision of business and strategic leadership. Examples include the Global Wind Energy Council and the World Wind Energy Association for wind power (there are also several national level wind energy associations), the European Photovoltaic Industry Association (which has an increasingly global focus), the International Hydropower Association, the World Bioenergy Association, and the International Geothermal Association. National industry groups are also becoming more sophisticated and prominent.

Renewable Energy Technologies

Most renewable energy resources originate from the sun (except for geothermal and some forms of ocean energy). Solar heat is the "motor" behind wind, biomass, and the water cycle tapped to produce hydropower; the earth core's nuclear processes and the gravity of the moon deliver the rest. In principle, the renewable energy resources are overly abundant. The sun provides the earth with radiation that in every hour provides roughly as much energy as the whole world consumes during a whole year. Yet many new technologies are being developed in order to convert the primary energy to secondary energy carriers that can be used such as electricity, heat (water, steam, air, etc), and fuels (see Table B.1).

All of these technologies use indigenous resources that are diversified all over the world. In some cases resources are available for free; however, the conversion technologies are not. One other common element of renewable energy is the dispersed quality compared to fossil or nuclear energy, which are quite concentrated. In general, harvesting renewable energy will consume space, preferable nearby the end consumer.

B.1. WIND ENERGY

Basically, a wind turbine transfers the kinetic energy of the wind through blades into alternating current (AC) power through a generator and (in most cases) a gear box. Two types of turbines exist: onshore and offshore. The latter is derived from the onshore turbine, although specific offshore turbine concepts have been developed and still are under development.

In general, offshore turbines are taller, their capacities are bigger, they have other foundations, and they have to meet higher demands concerning quality because of the harsh offshore conditions. Investments per megawatt for offshore are higher as well as operating and maintenance costs; however, offshore winds are generally stronger and more consistent than those on shore.

TABLE B.1   Renewable resources and secondary energy carriers

Source	Use in the form of
Wind	
Water pumping and battery charging	Mechanical energy, electricity
Onshore wind turbines	Electricity
Offshore wind turbines	Electricity
Biomass	
Combustion (domestic)	Heat (e.g., for cooking, space heating)
Combustion (industrial)	Process heat, steam, electricity
Gasification/power production	Power, heat, combined heat, and power/electricity.
Gasification/fuel production	Hydrocarbons, methanol, H2
Hydrolysis and fermentation	Ethanol
Pyrolysis/production of liquid fuels	Bio-oils
Pyrolysis/production of solid fuels	Charcoal
Extraction	Biodiesel
Digestion	Biogas
Solar	
PV solar energy conversion	Electricity
Solar thermal power	Heat, steam, electricity
Low-temperature solar energy	Water and space heating and cooling
Passive solar energy use	Heat, cold, light
Artificial photosynthesis	H2 or hydrogen rich fuels
Hydropower	
Mechanical energy, electricity	
Geothermal	
Heat, steam, electricity	
Marine	
Tidal energy	Electricity
Wave energy	Electricity
Current energy	Electricity
OTEC	Heat, electricity
Salinity gradient/osmotic energy	Electricity
Marine biomass production	Fuels

At the best sites, the costs of wind energy are directly competitive with fossil-fuel power production. For less favorable sites, wind energy can still compete if the environmental advantages are taken into account, for example, by eco-taxes or subsidies.

B.1.1. Trends

In 2011, the average capacity of wind turbines delivered to the market was 1.7 MW. The gradual increase of the average size creates economies of scale. The largest commercially available wind turbines are now in the 7.5 MW range, with prototypes up to 10 MW. Other trends in R&D are the reduction of weights and loads, the reduction of the number of components (e.g., the direct drive generator), improvements in grid integration and power quality (e.g., power electronics), and improved materials (e.g., for blades).

B.2. BIOMASS AND WASTE

A wide variety of biomass energy systems exist, using different types of biomass and producing various kinds of secondary energy carriers. In each biomass energy system, one can distinguish between the resources, logistics (production, preparation, transport), and biomass energy conversion.

B.2.1. Variety of Resources

Biomass is usually defined as all forms of plant-derived material that is part of a short carbon cycle (excluding fossilized material). As an energy source, it can be used in a renewable way if the rate of harvest is equal to or lower than the rate of regrowth. This makes energy from energy plantations and from residues of agriculture and sustainable forestry a form of renewable energy.

The basic principle behind the production of biomass is the process of photosynthesis—organic matter is produced by plants capturing CO_2 from the atmosphere. The harvested part of the carbon is re-emitted to the atmosphere during the conversion into secondary energy carriers. Therefore, the combination of photosynthesis and energy conversion is, in principle, CO_2-neutral, except for some fossil-fueled harvesting, processing, and transport.

There are many forms of biomass, ranging from very wet streams with a very low heating value, like animal manure, to very dry streams with a much higher heating value, like air-dried wood. Also, the chemical composition varies between different sources of biomass. As an energy source, biomass must compete with other applications (Figure B.1), except when it is cultivated as a dedicated energy crop (e.g., willows, miscanthus, rapeseed, grain).

B.2.2. How can Biomass be used?

The vast majority of biomass energy is applied as heat. In rural areas the use of (traditional) biomass for cooking or heating is still quite common. Modern use of biomass is shown, for example, in Scandinavian countries, where

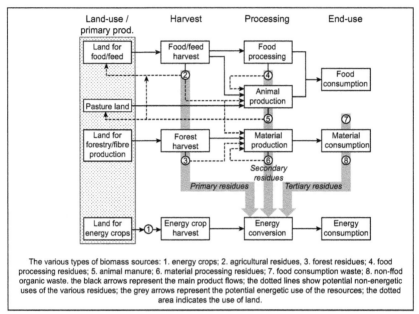

The various types of biomass sources: 1. energy crops; 2. agricultural residues, 3. forest residues; 4. food processing residues; 5. animal manure; 6. material processing residues; 7. food consumption waste; 8. non-ffod organic waste. the black arrows represent the main product flows; the dotted lines show potential non-energetic uses of the various residues; the grey arrows represent the potential energetic use of the resources; the dotted area indicates the use of land.

Figure B.1 Biomass sources.

district heating plants are the big consumers of biomass. At present, biomass electricity production is expected to grow faster than heat production.

Besides heat and electricity, there is also potential to use biomass in biofuels for transportation.

B.2.3. Logistics

Arising from the wide diversity of resources and conversion technologies, and of the low energy density of biomass (heating value 14 MJ/kg for air-dried wood, coal 24 MJ/kg, natural gas 38 MJ/kg, and fuel oil 40 MJ/kg), logistics (production, preparation, transport) are an important issue regarding biomass energy. Transportation is relatively expensive, which calls for size reduction, transport, drying, and storage (Figure B.2).

B.2.4. Conversion Technologies

Basically all types of energy carriers can be produced from biomass: electric energy, mechanical energy, heat, light, liquid and gaseous fuels, and

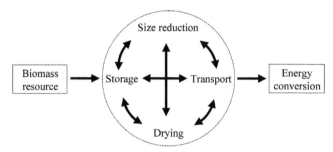

Figure B.2 Logistical steps within a biomass energy system.

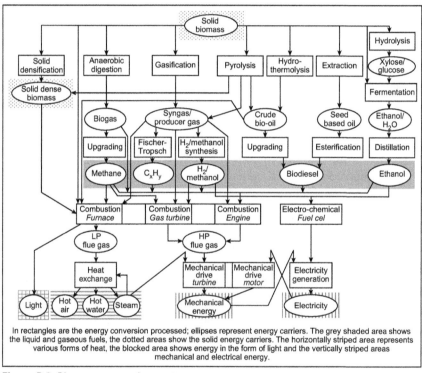

In rectangles are the energy conversion processed; ellipses represent energy carriers. The grey shaded area shows the liquid and gaseous fuels, the dotted areas show the solid energy carriers. The horizontally striped area represents various forms of heat, the blocked area shows energy in the form of light and the vertically striped areas mechanical and electrical energy.

Figure B.3 Biomass conversion routes.

convenient forms of solid fuels. The relevant conversion technologies (Figure B.3) can be grouped into four main categories:

- Thermochemical conversion by applying combustion (excess of oxygen), gasification (less oxygen), and pyrolysis and hydrothermolysis (no oxygen). From biomass, all kinds of products could be extracted, in bio–refineries or otherwise. In principle, products that have been extracted from fossil

fuels for ages (e.g., for the chemical industry) can be extracted from biomass, because they have a similar constitution of hydrocarbons. Such materials drawn from biomass could contribute to a "bio-based economy."
- Biochemical conversion: anaerobic digestion and fermentation.
- Extraction, for example, oil from rapeseed.
- Solid densification: such as in briquettes and pellets.

B.2.5. Trends in Technologies

One promising conversion technology for power plants is the cocombustion of biomass in coal-fired power generation. Its benefits are self-evident, such as economies of scale, due to the huge amounts of biomass that can be used. Cogasification is another opportunity, with the same economies of scale benefits. In addition, stand-alone biomass plants are more and more becoming viable.

The introduction of CO_2 emission standards for power plants (e.g., on the level of natural gas power plants) is believed to be a stimulating measure for cocombustion or cogasification, because of their ability to decrease CO_2 emissions from coal power plants substantially.

As for biofuels, woody fuels and algae ("second-" or "third-"generation biomass) warrant a lot of attention. These biomass resources would counter some serious problems with present biomass resources as palm oil or soy oil, such as a modest greenhouse gas balance, the competition with food production and land use change (resulting in possible deforestation). Woody fuels have a much better balance, but still need a technological breakthrough for an efficient conversion to biofuels for transportation. Aquatic algae have a high yield of bio-oil, but large-scale production of algae is still investigated. Fermentation processes are still a sound energy technology in all parts of the world, yielding biogas and subsequently heat for local use or for centralized distribution. This biogas can also be added to existing transport systems for natural gas.

B.3. SOLAR ENERGY

There are three types of technologies for tapping direct solar energy:
- Solar thermal heat systems
- Concentrating solar thermal power (CSP)
- Solar photovoltaics (PV)

B.3.1. Technologies

Direct use of solar heat is the predominant form of solar energy use, like in heating of houses. This energy use can be "passive" direct heating (through walls and windows), but also active, with the use of devices like solar boilers or storage systems. Most of this heat is used at the point of demand. See also Section B.8.

CSP is expected to become an important energy source in sunny, arid countries, for example, in the Mediterranean region in Europe and North Africa. In this technology solar radiation is used to heat water or other fluids to high temperatures (e.g., >100°C), where the heated steam or fluids can drive power generators. Concentration of solar radiation with mirrors is the common basic technology here.

PV conversion of sunlight still provides a tiny share of global electricity, but relative growth in capacity and generation has been rapid in recent years. This technology is based on semiconductors in which electrons are mobilized by solar irradiation, creating an electric potential and hence electric power. The efficiency of transposing solar radiation into electric energy in such solar cells is typically in the order of 10–20% (in the case of silicon), but can reach also almost 40% in experimental devices with rare materials.

B.3.2. Trends in Technologies

Mass-produced solar cells are traditionally semiconductor devices made of silicon, which is the most abundant element in the earth's crust. R&D focuses on scaling up the mass production technologies for multi- and single-crystalline solar modules (the majority of the market) and on improving thin film silicon cells and production technology (amorphous silicon, CdTe and CIS). The focus is also on the development of next-generation solar cells, such as dye-sensitized and organic solar cells. While these promise to be a lot less expensive, they are still some years away from commercial production.

In CSP, parabolic trough plants are the main technology, using parabolic mirrors to reflect the radiation to linear receivers. There is also a growing interest in other CSP technologies. Central receiver plants have one or more central receivers instead of linear receivers; Fresnel plants use flat Fresnel mirrors instead of parabolic mirrors, due to the lower cost for construction.

B.4. HYDROPOWER

Hydropower—the conversion of potential energy of water into secondary energy carriers by means of hydraulic systems—throughout history has played an important role in providing first mechanical, then electrical power. Of the various forms of renewable energy, hydropower is the single largest contributor, representing almost 75% of total global renewable electricity production. However, (large) hydro has a rather poor image with governments and agencies, is often excluded from (voluntary) green energy schemes, and is therefore not enjoying the boom in increased deployment like some other renewable sources have enjoyed in the last decade. The main reasons for this are the sometimes large environmental effects and social impacts of large hydro. Nevertheless, large hydro is still being built, especially in Asia and South America, while other markets are focusing on modernizing and refurbishing existing installations.

B.4.1. Technologies

Hydropower schemes are usually divided into two categories: large-scale (above 10MW) and small-scale (below 10 MW, including mini- and micropower schemes).

Most of the effort in developing hydropower has focused on the exploitation of "heads" (elevation) of 5 meter or more (often much more). Locations with such large falls of water represent the most highly concentrated form of renewable energy resource. If such a head is not available naturally, it may be created artificially by cutting streams and building a reservoir. This involves substantial modification of the local environment and high cost for the civil works.

A great deal of the hydroelectric resources in the world are used as energy storage devices. Such "pumped storage" capacity is independent of the original form of energy (fossil, nuclear, or renewable) and is usually based on reservoirs at different heights. Pure pumped storage capcity is excluded from hydro overviews in Chapter 1. When available, excess energy from the grid is used to pump water from the lower to the higher reservoir. In times of high electricity demand, water flows from the upper reservoir and is converted to electricity as in a conventional scheme.

Pumped storage is used to store surplus energy from conventional fossil-fueled plants and nuclear power plants, and, in the future, will also be applied to store renewable energy. The benefits of such systems can include rapid peak load response and stabilization of the network. Pure pumped storage capacity is not included as a renewable source in the market overviews in Chapter 1.

B.4.2. Trends in Technologies

Generally speaking, the most attractive sites in all countries have already been developed. Although large hydro is still being built, there is growing interest in small-scale and lower head sites, and in retrofitting and upgrading existing sites. Lower head sites are statistically much more common as there are many thousands of kilometers of rivers with low gradients. Reflecting this situation, in recent years considerable research effort has gone into the development of more efficient and economic technologies to exploit low-head hydro.

Furthermore, there is a growing international interest in the development of small-scale hydropower (SHP) because it does not have the same kind of adverse effects on the local environment as large hydro. SHP is, in most cases, "run-of-river", in other words any dam or barrage is quite small, and generally little or no water is stored. The engineering work involved merely serves to regulate the level of the water at the intake to the hydro-plant.

B.5. OCEAN ENERGIES

The oceans cover more than two-thirds of the surface of our planet and represent a potential, chemical, and kinetic energy resource, which theoretically is far larger than the entire human race could possibly use. The huge size of the marine energy resource is to some extent academic, as most of the energy available is either too diffuse for economic exploitation or located too far from the end-use sites. However, there are places where the different types of marine energy tend to be concentrated that may be located a "feasible" distance from prospective markets. In such cases, prospects for future exploitation are good.

There are six quite different marine energy resources that could be developed, as follows:
- Tidal and marine currents
- Wave energy
- Ocean thermal energy conversion (OTEC).
- Tidal barrages
- Salinity gradient/osmotic energy
- In addition, the ocean can also serve as a source for marine biomass, from which fuels can be produced. This type of ocean energy is categorized as a biomass resource.

B.5.1. Tidal and Marine Current Energy

Marine currents are mainly driven by the rise and fall of the tides, but also by differences in seawater composition and by oceanic circulation. The mechanisms for exploiting this kinetic energy resource are similar to those for wind energy. An advantage of the marine energy resource is that it is generally predictable since the drivers tend to be gravitational rather than meteorological.

In most places, the movement of seawater is too slow and the energy available is too diffuse to permit practical energy exploitation. However there are locations where water velocity is speeded up by a reduction in cross section of the flow area, such as straits between islands and the mainland, around the ends of headlands, in estuaries, and so forth. The main siting requirement is a location having flows exceeding about 1.5 m/s for a reasonable period with sufficient depth of water to cover a reasonable size of turbine (perhaps 15–30 meter).

Notable technological developments have taken place in the United States, Canada, the UK, Australia, and Japan.

The various turbine rotor options can generally be categorized as those devices that rely mainly on drag forces and those that rely predominantly on lift forces, the latter category is most promising due to its inherent higher efficiency.

Examples of drag devices include the traditional waterwheel and the Savonius type rotor. Lift devices can be classified by horizontal axis rotors and vertical axis rotors (similar to the Darrieus type rotor).

B.5.2. Wave Energy

Ocean waves are caused by winds as they blow across the surface of the sea. The energy that waves contain can be harnessed and used to produce electricity. Due to the direction of the prevailing winds and the size of the Atlantic Ocean, the UK and northwestern Europe have one of the largest wave energy resources in the world.

B.6. GEOTHERMAL ENERGY FOR POWER

Geothermal energy is generally defined as heat stored within the earth. This heat originates from the earth's molten interior and from the decay of radioactive materials. Down to the depths accessible by drilling with modern technology, the average geothermal gradient is about 2.5–3.0 °C/100 m. Geothermal energy can be used for electricity generation or as a direct heat source. In some countries, such as Italy, harvesting power and heat from geothermal sources has been quite common for many years. About two-thirds of all geothermal commercial capacity is used for direct heat (see Section B.7).

B.6.1. Trends in Technologies

Geothermal electricity generation capacity is growing at a modest rate. However, geothermal power is becoming more attractive because new technologies (such a low-temperature bottoming cycles or reservoir enhancement) are making plants more efficient. Moreover, small-scale geothermal power plants are becoming more and more "turn key." Next-generation microturbine technologies using geothermal heat are being demonstrated.

B.7. RENEWABLE HEATING AND COOLING TECHNOLOGIES

Renewable energy sources used for heating and cooling include solar radiation, geothermal energy, and biomass (in solid, liquid, or gaseous forms and derived from a variety of feedstocks). These technologies can be implemented in large-scale, centralized applications such as district heating systems or combined heat and power (CHP) plants, as well as small-scale, decentralized applications such as individual households. This appendix provides a more in-depth introduction to the solar, geothermal, and biomass technologies available for heating and cooling applications.

In district heating systems, hot water or steam is distributed to buildings or industries via an insulated two-pipe network. Heat exchangers in individual buildings act as an interface, capturing and circulating heat internally—returning cooler water to the network. Cold water or slurries are increasingly used in district heating networks to provide cooling services. District heating networks have most often been implemented in densely populated areas with strong planning powers, where a centralized body plans and builds the necessary infrastructure.

Such networks serve a majority of citizens in several countries such as Iceland (99%), Latvia (64%), and Denmark (61%). Depending on local conditions and availability, it is typically possible to integrate high shares of renewable energies such as biomass and geothermal into existing large-scale applications such as fossil fuel-based district heating networks without technical problems because of their high capacity factors.

In a similar large-scale application, renewable sources can be used to simultaneously produce heat and electricity in CHP plants. The use of CHP is common in industries with large concurrent heat and power demands, and has been increasing in medium-scale applications as well, such as in the food processing, ceramics, textiles, and pulp and paper industries. The renewable technologies that may be applied to these systems are discussed in the following large- and medium-scale applications section.

B.7.1. Solar Thermal

In solar heating systems, collectors transform solar irradiance into heat and use a carrier fluid (e.g., water or air) to transfer the heat either to point of use or to a storage unit. There are a number of different technologies, both active (e.g., flat plate collectors and evacuated tube collectors) and passive (e.g., integral collector storage and thermosyphon systems) that use collectors.

Large- and Medium-Scale Solar Thermal Applications

There are several examples of solar thermal collectors that have been integrated into existing district heating systems. However, storage requirements increase with increasing percentages or shares of solar thermal in the system's heat supply and as a result, use of solar thermal technologies in district heating systems is much less common than geothermal or bioenergy.

Concentrating solar heating systems can be used to provide higher grade heat (up to 400°C) for industry, agriculture, and food production. These systems use a concentrator to capture solar irradiation and then direct it to a receiver where the heat energy is absorbed by a carrier fluid (normally a special type of oil). This fluid is transported for direct use via a heat exchanger or stored.

Solar thermal panels producing medium-temperature heat (up to 150°C) are in the early stages of development and may provide useful energy for industry in the future. Worldwide, there are currently less than 100 medium-temperature solar panels in operation.

Although less common than heating systems, solar thermal-based cooling systems with capacities up to several hundred kilowatts have been built. Such systems can use evacuated tubes to drive absorption cycles, thereby providing cooling services.

Small-Scale Solar Thermal Applications

The most common application of solar thermal heat is in the form of low-temperature (up to 80°C) collectors that provide heat to residential or commercial buildings. Flat plate collectors are most widely used at the residential level, providing space or hot water heating. Small solar-thermal based cooling systems (from 100 down to 4.5 kW) have also been developed in recent years.

B.7.2. Geothermal

Geothermal technologies for heating purposes may be broken down into two categories: direct-use and heat pumps. Direct-use geothermal technologies provide space heating by means of open (single-pipe) or closed (double-pipe) loop systems, both of which use heated geothermal water from wells, and circulate spent water back into injection wells. Heat pumps upgrade low-temperature ambient heat (e.g., from air, water, or ground) with the assistance of electricity to create useful higher temperature heat that can be used for heating systems up to 45°C, or to provide cooling.

Large- and Medium-Scale Geothermal Applications

Direct-use geothermal technologies are mature and can provide space heating for buildings, including via district heating systems, but can also provide heat for industrial processes, for example, agricultural products and mineral drying. Direct-use geothermal can also supply CHP plants with both electricity and heat. Iceland, for example, as of 2010 had a total of 580 MW$_{th}$ of geothermal-based CHP capacity installed.

Small-Scale Geothermal Applications

Direct-use geothermal technologies can also be deployed at smaller scales, for example, in individual residences. Heat pumps are most often associated with small-scale, decentralized applications. At the beginning of 2010 around three million heat pumps were installed globally, producing 5 Mtoe of geothermal based heat.

B.7.3. Bioenergy

Bioenergy heating and cooling technologies convert the stored solar energy in biomass into usable forms. Feedstocks can originate as agricultural products (e.g., energy crops), forest products (e.g., wood residues), waste (e.g., animal waste or municipal solid waste), or industrial byproducts (e.g., lignin-containing sulfite lyes in alkaline-spent liquor from pulp and paper production). The generation of bioenergy-based heat can involve complex pretreatment, upgrading, and conversion processes (see Figure B.4).

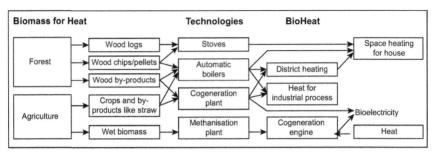

Figure B.4 Possible conversion routes for biomass heat (simplified) (*EREC, 2007*).

Large- and Medium-Scale Bioenergy Applications

Biomass heat is common in large-scale, industrial boilers and can also be used in district heating applications. Biomass feedstocks such as woody biomass, crop residues, pellets, and solid organic wastes are often used in combination with conventional fuels in district heating systems. Large-scale boilers are able to burn a larger variety of fuels, including wood waste and refuse-derived fuel, whereas medium-scale systems typically use wood chips. Forestry and agriculture residues as well as the biogenic component of municipal residues and wastes are often used in CHP production and can reach overall conversion efficiencies of around 70–90%.

In addition, bioenergy feedstocks (e.g., from landfills, urban sewage, and industrial and agricultural wastes) can be used in anaerobic digestion processes to produce biogas. Until recently, the most common application of biogas has been in onsite heat production or heat production via dedicated local systems. It is, however, becoming increasingly common to upgrade biogas to biomethane of a quality suitable for blending into local, regional, or national natural gas pipelines.

Small-Scale Bioenergy Applications

Biomass heat production is also common in small, individual household systems. Examples of small-scale technologies using biomass feedstocks for heat production include wood burning stoves, small-scale boilers, and pellet-based domestic heating systems. Small-scale household systems typically use wood logs or pellets as fuel.

B.8. RENEWABLE TRANSPORTATION TECHNOLOGIES

Renewable energy technologies can be applied at different places in the transportation supply chain (from primary source to type of vehicle; see Figure B.5).

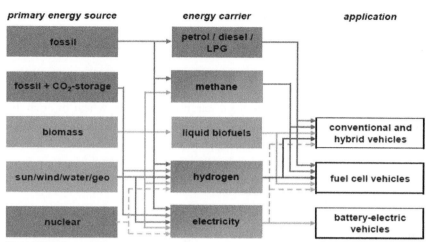

Figure B.5 From primary energy source to end use in vehicles. The lighter gray regions represent the technologies where renewable energy can be applied *(From RETRANS)*.

B.8.1. Biofuels

Biodiesel

The most commonly produced liquid biofuels for transportation are biodiesel and ethanol. Chemical and energy characteristics of commercial biodiesel are largely the same as fossil diesel. There are slight differences between biodiesel (or fatty-ester methyl ether; FAME), pure plant oil, and synthetic biodiesel.

Biofuel blends can be burned in conventional combustion engines with or without adaptations, depending on the type of fuel used. Low-percentage blends (5–20%) in conventional diesel can be used in conventional engines. Pure biofuels require engine adaptations to specific characteristics of the fuel. Some biofuels even enable engine adaptations that improve performance or energy efficiency.

Biogas is also expected to be used in road transport, mainly in public transport (city buses) and municipal fleets. Biomethane is already in use in some European countries, fed directly into the natural gas grid (for heat and power generation and for fueling vehicles).

So-called first-generation biodiesels are commonly produced from vegetable oils, extracted from oil palm, rapeseed, sunflowers, soybeans, and so forth. The vegetable oils are converted through transesterification into FAME, which can be used in place of petroleum diesel.

Bio-Ethanol

Ethanol can substitute gasoline. Bio-ethanol (produced from biomass) is commonly used in low blends in gasoline, typically 5% (E5) or 10%

(E10) on volume basis. Higher blends can be used in flexible fuel vehicles (up to E85), which are currently offered by a wide range of manufacturers.

First-generation bio-ethanol is produced by fermentation of sugars. These sugars can be extracted from feedstocks like sugar beet and sugar cane, or the sugars can be made from starch in crops like wheat or corn. In particular, shipping also offers an opportunity to use cheaper, low-quality biofuels, because marine engines are typically suited to use a low fuel quality (fuel oil). But this opportunity has hardly been applied yet.

Second Generation
In addition, several types of biofuels are currently under development that could become available in the mid or long term. Most of these technologies use lignocellulosic biomass ("second generation") in the form of wood residues, paper waste, agricultural waste, and dedicated energy crops. Feedstocks of this type are far larger and are expected to be more "sustainable," because of higher yields, lower net greenhouse gas emissions, and a lack of competition with food production.

Currently, creating biofuels from algae (hydro-cultures that could provide high yields of oil) is gaining interest. Test flights of planes with biofuels (in at least one case originating from algae) have been performed.

B.8.2. Hydrogen
Hydrogen is a versatile energy carrier that can be produced from any fossil or nonfossil primary energy source. It can be converted into mechanical energy in combustion engines and into power using so-called "fuel cells,", but the business case has not been widely spread yet. Although (fossil) hydrogen was already applied in some plane flights, technical issues like on-board storage still have to be solved. Technologies are still in their early days of reducing costs—even more if combined with renewable sources—and overcoming technical barriers.

B.8.3. Electric Vehicles
Vehicles running on electricity do not necessarily fall in the category of sustainable transport, because they may still use power from fossil or nuclear power plants. Nevertheless, if the share of power from renewable sources increases, electric vehicles (EVs) also become more sustainable. In some cases, EVs could exclusively run on power from renewable sources, for example, when the batteries are charged in a solar-powered filling station.

In general, two types of EVs are being developed. Full-electric vehicles are essentially powered by the grid, by means of storing power in batteries. Trains, subways, trams, and trolley buses are grid-connected and can also be regarded as full-electric transportation.

So-called "hybrids" combine an electric drive system with a conventional engine and batteries; they run on electricity and fuel. Hybrids do not necessarily connect to the grid; special "plug-in" hybrids provide that opportunity.

Full-Electric Vehicles

The concept of EVs is actually quite old, dating back to around 1900. Although fuels won the battle in those days, EVs are returning. Especially driven for their local benefits on the environment, many regions and cities are setting up trials and programs with EVs.

Batteries will be crucial for their success. Battery capacity, lifetime, safety in automotive applications, charging, and system design are all issues that are investigated. Concerning safety fire hazards, associated with the use of lithium, need to be minimized.

New concepts for speed-charging or exchange of batteries may be boosting the market, because they increase the range. Different vehicles require different battery types. Light-weight EVs require batteries with a high energy density (kWh/kg).

EVs are expensive, but that is partly compensated by low energy costs. Due to high petrol taxes, operating costs per mile or kilometer of an EV are relatively low.

An important issue in the future will be the interconnection of large quantities of EVs to the grid. Charging of batteries in off-peak hours could improve the economic efficiency of the power system. Basically, car batteries could even deliver to the grid during peak demand hours, which also improves the efficiency. Regarding coevolution of EVs and renewable energy, see Chapter 8.

Hybrid Vehicles

Hybrid vehicles contain two engines: (1) one electric motor, which uses the power to drive the vehicle and (2) one combustion engine (or a fuel cell), which supplies electric power to the electric motor or the batteries. In "parallel" systems the combustion engine also drives the vehicle.

Hybrids generally require smaller batteries than full-electric cars, because they also carry fuel. The energy content may be limited but the specific power (kW/kg) has to be high.

Hybrid vehicles only turn "renewable" if they use fuel (diesel, petrol, hydrogen) or power from renewable sources.

References

Alagappan, L., Orans, R., Woo, C.K., 2011. What drives renewable energy development? Energy Policy 39, 5099–5104.

Alonso, O., Revuelta, J., de la Torre, M., Coronado, L., 2008. Spanish experience in wind energy integration. In: Power-Gen Conference and Exhibition, Milan, Italy, pp. 3–5.

Altenburg, T., Schmitz, H., Stamm, A., 2008. Breakthrough? China's and India's transition from production to innovation. World Development 36 (2), 325–344.

Arvizu, D., Balaya, P., Cabeza, L., Hollands, T., Jäger–Waldau, A., Kondo, M., et al., 2011. Direct Solar Energy. In: Edenhofer, O., Pichs-Madruga, R., Sokona, Y., Seyboth, K., Matschoss, P., Kadner, S. (Eds.), IPCC Special Report on Renewable Energy Sources and Climate Change Mitigation, Cambridge University Press, Cambridge.

Asia Development Bank: http://www.adb.org/Media/Articles/2011/13530-india-solar-power/ April 2011

Beck, F., Martinot, Eric, 2004. Renewable Energy Policies and Barriers, Encyclopedia of Energy. In: Cleveland, Cutler J. (Ed.), Academic Press/Elsevier Science.

Beurskens, L., Hekkenberg, M., 2011. Renewable Energy Projections as Published in the National Renewable Energy Action Plans of the European Member States. Energy Research Center of the Netherlands.

Bhattacharya, S., Jana, C., 2009. Renewable energy in India: Historical developments and prospects. Energy 34, 981–991.

BNEF, 2011. Clean energy investment 2010-fact pack presentation. http://bnef.com/Press Releases/view/138 Press release 25-1-2011.

Bolinger, M., Porter, K., 2002. Renewable Energy Loan Programs Case Studies of State Support for Renewable Energy. Berkeley Lab and the Clean Energy Group.

Bomb, C., McCormick, K., Deurwaarder, E., Kaberger, T., 2007. Biofuels for transport in Europe: lessons from Germany and the UK. Energy Policy 35 (2007), 2256–2267.

Breitbart: ADB to launch pension-related fund for Asia clean energy projects, http://www.breitbart.com/article.php?id=D9LETNJ80&show_article=1, February 18, 2011.

Brown Rudnick, 2011. Connecticut Energy Reform Legislation Summary of SB 1243. http://www.brownrudnick.com/nr/pdf/alerts/Brown_Rudnick_Connecticut_Energy_Reform_Legislation__Summary_of_SB_1243_6-11.pdf.

BTM consult, 2011. BTM consult releases new wind report: World Market Update 2010. http://www.btm.dk/public/Navigant_WMU2010_ReportRelease.pdf, Washington, DC: 28 March 2011.

Bundesministerium für Umwelt, Naturschultz und Reaktorsicherheit (BMU), 2009. Renewable Energy Sources in Figures: States. National and International Development, Berlin, Germany 80 pp.

Bundesministerium für Umwelt, Naturschultz und Reaktorsicherheit (BMU), 2010. Zeutreihen zur Entwicklung der erneuerbaren Energien in Deutschland [Development of Renewable Energy Sources in Germany], Berlin, Germany.

Bürer, M.J., Wüstenhagen, Rolf, 2009. Which renewable energy policy is a venture capitalist's best friend? Empirical evidence from a survey of international cleantech investors. Energy Policy 37 (2009), 4997–5006.

Bürger, V., Klinski, S., Lehr, U., Leprich, U., Nast, M., Ragwitz, M., 2008. Policies to support renewable energies in the heat market. Energy Policy 36 (8), 3150–3159.

Center for American Progress, China's Clean Energy Push (June 21, 2010). http://www.americanprogress.org/issues/2010/06/pdf/china_hill_memo.pdf.

CITI, 2010. Europe Utilities (Citigroup Global Markets), September 2010. Citi Investment Research & Analysis, London, UK.

Clean Energy Authority: http://www.cleanenergyauthority.com/solar-installers/.

Clean Energy Group: http://www.cleanegroup.org/blog/open-those-energy-innovation-doors/

Climate Bonds Initiative: www.climatebonds.net.

Climate Progress, 2011. One third of the World's energy could be solar by 2060, 2–12-2011.

Connor, P., Bürger, V., Beurskens, L., Ericsson, K., Egger, C., 2009. Overview of RESH/RES-C Support Options. D4 of WP2 from the RES-H Policy project. Exeter, UK. University of Exeter, Available at www.res-h-policy.eu/downloads/RES-H_Policy-Options_(D4)_final.pdf.

Couture, T., 2009. State Clean Energy Policy Analysis: Renewable Energy Feed-in Tariffs. SCEPA Webinar. National Renewable Energy Laboratory, Golden, CO, USA 23 pp.

Couture, Toby D., Cory, Karlynn, Kreycik, Claire, Williams, Emily, 2010. A Policymaker's Guide to Feed-in Tariff Policy Design. Technical Report NREL/TP-6A2–44849 National Renewable Energy Laboratory, July.

Cruetzig, F.S., Kammen, D.M., 2010. Getting the carbon out of transportation fuels. In: Nobel Cause, A., Schellenhuber, H.-J., Molina, M., Stern, N., Huber, V., Kadner, S. (Eds.), Global Sustainability, Cambridge University Press, pp. 307–318.

Dalenbäck, J.O., 2010. Success factors in solar district heating. CIT Energy management AB.

Damborg, S., Krohn, S., 1998. Public Attitudes towards Wind Power. Danish Wind Turbine Manufacturers Association, Copenhagen, Denmark.

De Gorter, H., Just, D.R., 2010. The Social costs and benefits of biofuels: The intersection of environmental, energy and agricultural policy. Applied Economic Perspectives and Policy 32 (1), 4–32.

De Jager, D., Klessmann, C., Stricker, E., Winkel, T., de Visser, E., Koper, M., Ragwitz, M., Held, A., Resch, G., Busch, S., Panzer, C., Gazzo, A., Roulleau, T., Gousseland, P., Henriet, M., Bouille, A., 2011. Financing Renewable Energy in the European Energy Market, prepared for the European Commission. Ecofys.

DECC, 2011. Renewable Heat Incentive. United Kingdom: Department of Energy and Climate Change.

Doris, E., McLaren, J., Healey, V., Hockett, S., 2009. State of the States 2009: Renewable Energy Development and the Role of Policy. NREL/TP-6A-46667 (Golden, CO: U.S. National Renewable Energy Laboratory, October 2009).

DSIRE, 2011. US Database of State Incentives for Renewables and Efficiency. Online database, US Database of State Incentives for Renewables and Efficiency (DSIRE). Available at North Carolina State University, Raleigh, NC, USA. www.dsireusa.org/.

DUKES, 2009. Digest of United Kingdom Energy Statistics (DUKES). Department of Energy and Climate Change, London, UK. Available at: www.decc.gov.uk/en/content/cms/statistics/publications/dukes/dukes.aspx. www.decc.gov.uk/en/content/cms/statistics/publications/dukes/dukes.aspx.

Eco Periodicals, July 2010.

Russell, Edward, 2011. Google Renewables. http://www.projectfinancemagazine.com/Article/2847544/Google-renewable.html 2011 June 14, 2011.

EPIA Global Market Outlook. May 2012.

Epstein, P. R., Buonocore, J. J., Eckerle, K., Hendryx, M., Stout III, B. M., Heinberg, R., Clapp, R. W., May, B., Reinhart, N. L., Ahern, M. M., Doshi, S. K., Glustrom, L., 2011. Full cost accounting for the life cycle of coal. (pages 73–98), February 2011.

EREC, 2007. Renewable Heating Action Plan for Europe. European Renewable Energy Council, Brussels.

Espey, S., 2001. Renewables portfolio standard: a means for trade with electricity from renewable energy sources? Energy Policy 29 (7), 557–566.

ESTIF, 2007. Best practice regulations for solar thermal. European Solar Thermal Industry Federation.

Euroheat and Power, 2010. District Heating and Cooling – 2009 Statistics. Euroheat and Power, Brussels, Belgium.

European Commission, 2005. The Support of Renewable Energy Sources. European Commission, Brussels, Belgium.

European Commission, 2011. Commission staff working document: Recent progress in developing renewable energy sources and technical evaluation of the use of biofuels and other renewable fuels in transport in accordance with Article 3 of Directive 2001/77/EC and Article 4(2) of Directive 2003/30/EC. Accompanying document to the Communication from the Commission to the European Parliament and the Council Renewable Energy: Progressing towards the 2020 target; COM(2011) 31 final. European Commission, Brussels, Belgium.

European Commission Cordis: http://cordis.europa.eu/fp7/what_en.html.

European Commission: http://www.sme-finance-day.eu/?id=8&print=1&no_cache=1&L=0.

European Parliament, 2009, April 23. Directive 2009/28/EC Of The European Parliament and of the Council. Official Journal of the European Union.

European Photovoltaic Industry Association (EPIA), 2011. Global Market Outlook for Photovoltaics Until 2015. Brussels.

European Photovoltaic Industry Association, Market Report 2011. January 2012. Brussels.

Farrell, J., 2009. Feed-in tariffs in America: Driving the Economy with Renewable Energy Policy that Works. The New Rules Project, Minneapolis, MN, USA 30 pp.

Felix-Saul, R., 2008. Assessing the impact of Mexico's Biofuels Law. Baker & McKenzie Biomass Magazine. Available at, http://biomassmagazine.com/articles/1678/assessing-the-impact-of-mexico's-biofuels-law2008.

Forsyth, T.L., Pedden, M., Gagliano, T., 2002. The Effects of Net Metering on the Use of Small-Scale Wind Systems in the United States. NREL/TP-500–32471 NREL, Golden, CO, USA 20 pp.

Frondel, Manuel, Ritter, Nolan, Schmidt, Christoph M, Vance, Colin, 2010. Economic impacts from the promotion of renewable energy technologies: the German experience. Energy Policy 38, 4048–4056.

Fulton, M., Kahn, B. M., Mellquist, N., Soong, E., Baker, J., Cotter, L., 2009. Paying for Renewable Energy: TLC at the Right Price. Achieving Scale Through Efficient Policy Design (Deutsche Bank Climate Change Advisors, December 2009).

Girardet, H., Mendonca, M., 2009. A Renewable World: Energy, Ecology, Equality. Green Books, Devon, UK.

Global Wind Energy Council (GWEC), 2011. Global Wind Report: Annual Market Update 2010. Brussels.

Global Wind Energy Council, Global Wind Statistics 2011, Brussels, 7 February 2012.

Goldstein, B., Hiriart, G., Bertani, R., Bromley, C., Gutiérrez–Negrín, L., Huenges, E., et al., 2011. Geothermal Energy. In: Edenhofer, O., Pichs-Madruga, R., Sokona, Y., Seyboth, K., Matschoss, P., Kadner, S. (Eds.), IPCC Special Report on Renewable Energy Sources and Climate Change Mitigation, Cambridge University Press, Cambridge.

Green Alliance, 2009. Establishing a Green Investment Bank for the UK. at 6.

Greenglass, N., Smith, K., 2006. Current improved cookstove acitivities in South Asia, a web-based survey. Woods Hole Research Center.

Greenpeace International: http://www.greenpeace.org/international/en/publications/reports/energy-revolution-a-sustainab/.

GWEC, 2011. GWEC Global Wind Report; Annual Update 2011.

Haas, R., Eichhammer, W., Huber, C., Langniss, O., Lorenzoni, A., Madlener, R., Menanteau, P., Morthorst, P.-E., Martins, A., Oniszk, A., Schleich, J., Smith, A., Vass, Z., Verbruggen, A., 2004. How to promote renewable energy systems successfully and effectively. Energy Policy 32, 833–839.

Haas, Reinhard, Christian Panzer, Gustav Resch, Mario Ragwitz, Gemma Reece, Anne Held, 2011. A historical review of promotion strategies for electricity from renewable energy sources in EU countries. Renewable and Sustainable Energy Reviews 15, 1003–1034.

Han, J., Mol, A., Lu, Y., Zhang, L., 2009. Onshore wind power development in China: Challenges behind a successful story. Energy Policy 37 (8), 2941–2951.

Held, A., Ragwitz, M., Huber, C., Resch, G., Faber, T., Vertin, K., 2007. Feed-in Systems in Germany, Spain and Slovenia. A Comparison. Fraunhofer Institute Systems and Innovation Research, Karlsruhe, Germany.

Hertel, T.W., Golub, A.A., Jones, A.D., O'Hare, M., Plevin, R.J., Kammen, D.M., 2010. Effects of US maize ethanol on global land use and greenhouse gas emissions: Estimating market-mediated responses. BioScience 60, 223–231.

HM Government, Update on the design of the Green Investment Bank.

Hofman, Daan, Huisman, Ronald, 2011. Did the financial crisis lead to changes in private equity investor preferences regarding renewable energy policies? Erasmus School of Economics. 2011. Available at Erasmus University, Rotterdam, Netherlands. http://people.few.eur.nl/rhuisman/images/PE%20investors.pdf2011.

Hogan, M., 2009. German biodiesel firms say U.S. imports escape duty. Reuters, 30 Nov 2009. Available at www.reuters.com/article/2009/11/30/us-germany-biodiesel-usidUSTRE5AT3QG200911302009.

Hunton, Williams, 2010. South Korea's Low Carbon. Green Growth Initiativehttp://www.hunton.com/emailblast/pdfs/renewable_energy_quarterly_february_2010.pdf.

Hvelplund, F., 2006. Renewable energy and the need for local energy markets. Energy 31 (13), 2293–2302.

IEA, 2001. Solar Energy Perspectives.

IEA-RETD, 2006. Renewable Energy Technology Deployment (RETD) - Barriers. Challenges and Opportunities.

IEA and IEA-RETD, 2008. Renewables for Heating and Cooling – Untapped Potential.

IEA-RETD, 2008. Policy instrument design to reduce financing costs in renewable energy technology projects.

IEA-RETD, 2010. Best Practices in the Deployment of Renewable Energy for Heating and Cooling in the Residential Sector.

IEA-RETD, 2011. Offshore Renewable Energy—Accelerating the deployment of offshore wind. tidal and wave technologies.

IEA-RETD, 2011. Renewable Energy for Residential Heating and Cooling—Policy Handbook. Earthscan, London.

IEA-RETD, 2011. Scoping Study—Linking RE Promotion Policies with International Carbon Trade (LINK).

IEA-RETD, 2012. Optimised use of renewable energy through improved system design (OPTIMUM).

Inderst, G., 2009. Pension Fund Investment in Infrastructure. OECD Working Papers on Insurance and Private Pensions No. 32.

International Energy Agency, 2007. Available at: http://www.iea.org/impagr/cip/pdf/issue 45hjkinterview.pdf.

International Energy Agency, 2008. Deploying Renewables. Principles for Effective Policies, Paris, France 200 pp.

International Energy Agency, 2009. Energy Technology Transitions for Industry. OECD/IEA, Paris.

International Energy Agency, 2010. Global Renewable Energy Policies and Measures Database. OECD/IEA, Paris.

International Energy Agency, 2010. World Energy Outlook 2010. OECD/IEA, Paris.

International Energy Agency, 2011. Clean Energy Progress Report. IEA Input to the Clean Energy Ministerial, Paris.

International Energy Agency, 2011. IEA analysis of fossil-fuel subsidies. Available at http://www.iea.org/weo/Files/ff_subsidies_slides.pdf2011.

International Energy Agency, Philibert, C, 2006. Barriers to Technology Diffusion. The Case of Solar Thermal Technologies, Paris.

IPCC, 2011. IPCC Special Report on Renewable Energy Sources and ClimateChange Mitigation. In: Edenhofer, O., Pichs–Madruga, R., Sokona, Y., Seyboth, K., Matschoss, P., Kadner, S., Zwickel, T., Eickemeier, P., Hansen, G., Schlömer, S., von Stechow, C. (Eds.), Cambridge University Press, Cambridge, UK and New York.

ITIF, 2009. Rising Tigers, Sleeping Giant. November 2009. Available at Breakthrough Institute, http://thebreakthrough.org/blog/Rising_Tigers_Summary.pdf2009.

Jaccard, M., Melton, N., Nyboer, J., 2011. Institutions and processes for scaling up renewables: Run-of-river hydropower in British Columbia. Energy Policy. http://dx.doi.org/10.10 16/j.enpol.2011.02.035.

Jacobsson, S., Bergek, A., Finon, D., Lauber, V., Mitchell, C., Toke, D., Verbruggen, A., 2009. EU renewable energy support policy: Faith or facts? Energy Policy 37 (6), 2143–2146.

Joanna Lewis, Ryan Wiser, 2005. Fostering a Renewable Energy Technology Industry: An International Comparison of Wind Industry Policy Support Mechanisms. (Lawrence Berkeley National Labs, November 2005). Available at http://eetd.lbl.gov/ea/emp/reports/59116.pdf.

Junginger, M., van Dam, J., Zarrilli, S., Ali Mohamed, F., Marchal, D., Faaij, A., 2011. Opportunities and barriers for global bioenergy trade. Energy Policy 39 (4), 2028–2042.

Katofsky, et al., 2010. Achieving Climate Stabilization in an Insecure World: Does Renewable Energy Hold the Key?

KFW IPEX-Bank: http://www.kfw-ipex-bank.de/ipex/en/Business_Sectors/Power_Rene wables_Water/Renewable_energy/index.jsp.

Klein, A., Held, A., Ragwitz, M., Resch, G., Faber, T., 2008. Evaluation of Different Feed-in Tariff Design Options – Best Practice Paper for the International Feed-in Cooperation. Fraunhofer Institute Systems and Innovation Research and Energy Economics Group, Karlsruhe, Germany and Vienna, Austria.

Klein, A., Merkel, E., Pfluger, B., Held, A., Ragwitz, M., Resch, G., Busch, S., 2010. Evaluation of Different Feed-in Tariff Design Options – Best Practice Paper for the International Feed. Cooperation, 3rd edition. Energy Economics Group and Fraunhofer Institute Systems and Innovation Research, Vienna, Austria and Karlsruhe, Germany.

Kolk, Ans, Pinkse, Jonatan, 2009. Business and climate change: key challenges in the face of policy uncertainty and economic recession. Management Online Review May 2009.

Kreycik, C., Couture, T. D., Cory, K. S., 2011. Innovative Feed-in Tariff Designs that Limit Policy Costs. Technical Report NREL/TP-6A20–50225 (Golden, CO: National Renewable Energy Laboratory, June 2011).

Lamers, P., Hamelinck, C., Junginger, M., Faaij, A., 2011. International bioenergy trade – a review of past developments in the liquid biofuels market. Renewable & Sustainable Energy Reviews 15 (6), 2655–2676.

Langniss, O., Seyboth, K., 2007. Experiences with the German Market Stimulation Program. 3rd European Solar Thermal Enegy Conference Proceedings, Freiburg.

Lantz, E., Doris, E., 2009. State Clean Energy Practices: Renewable Energy Rebates. NREL/TP-6A2–45039. National Renewable Energy Laboratory, Golden, CO, USA 38 pp.

Lapola Schaldach, D.M.R., Alcamo, J., Bondeau, A., Koch, J., Koelking, C., Priess, J.A., 2010. Indirect land use changes can overcome carbon savings from biofuels in Brazil. Proceedings of the National Academy of Sciences 107 (8), 3388–3393.

Levi, et al., 2009. Energy Innovation. (Council of Foreign Relations: November 2010) at 9.

Lewis, J., Wiser, R., 2005. Fostering a Renewable Energy Technology Industry: An International Comparison of Wind Industry Policy Support Mechanisms. LBNL-59116. Ernest Orlando Lawrence Berkeley National Laboratory, Berkeley, CA, USA 30 pp.

Liao, C., Jochem, E., Zhang, Y., Farid, N.R., 2010. Wind power development and policies in China. Renewable Energy 35, 1879–1886.

Lipp, J., 2007. Lessons for effective renewable electricity policy from Denmark, Germany and the United Kingdom. Energy Policy 35 (11), 5481–5495.

Lund, Henrik, 2007. Renewable energy strategies for sustainable development. Energy 32, 912–919.

Lund, Henrik, 2010. The implementation of renewable energy systems. Lessons learned from the Danish Case. Energy 35, 4003–4009.

Mendonça, M., 2007. Feed. Tariffs: Accelerating the Deployment of Renewable Energy, Earthscan, London, UK.

Mendonça, M., Lacey, S., Hvelplund, F., 2009. Stability, participation and transparency in renewable energy policy: Lessons from Denmark and the United States. Policy and Society 27, 379–398.

Metcalf, G.E., 2008. Tax Policy for Financing Alternative Energy Equipment. Tufts University Economics Department Working Paper series. Tufts University, Medford, MA, USA.

Mitchell, C., 2000. The Non-Fossil Fuel Obligation and its future. Annual Review of Energy and Environment 25, 285–312.

Morales, A., Robe, X., Sala, M., Prats, P., Aguerri, C., Torres, E., 2008. Advanced grid requirements for the integration of wind farms into the Spanish transmission system. Renewable Power Generation. IET 2 (1), 47–59.

Munich RE, 2001. Available at: http://www.munichre.com/en/media_relations/press_releases/2010/2010_12_23_press_release.aspx.

Musall, David, Fabian, Kuik, Onno, 2011. Local acceptance of renewable energy—A case study from southeast Germany. Energy Policy 39, 3252–3260.

Nast, M., Langniss, O., Leprich, U., 2007. Instruments to promote renewable energy in the German heat market—Renewable Heat Sources Act. Renewable Energy 32, 1127–1135.

Navigant Consulting, 2009. Ground–Source Heat Pumps: Overview of Market Status, Barriers to Adoption, and Options for Overcoming Barriers. U.S. Department of Energy.

Nelson, T., Simshauser, P., Kelley, S., 2011. Australian residential solar feed-in tariffs: industry stimulus or regressive form of taxation. Applied Economic and Policy Research, Working Paper No. 25-FIT, North Sydney.

Neuhoff, K., 2004. Large Scale Deployment of Renewables for Electricity Generation. Organisation for Economic Co-operation and Development, Paris, France.

Norad: http://www.norad.no/en/Thematic+areas/Energy/Clean+Energy/How+we+work/Partners/The+World+Bank+Group.146037.cms.

Pathways in Energy Technology Perspectives 2012. (figure 17).

Personal communication Kees van der Leun, Ecofys.

Pew Charitable Trusts, 2010. Who's Winning the Clean Energy Race. A G-20 Fact Sheet on Clean Energy Investment. Available at http://www.pewglobalwarming.org/cleanenergy economy/pdf/PewG-20Report.pdf.

Pieprzyk, Björn, Paul Rojas Hilje, 2009. Renewable Energy—Predictions and Reality: Comparison of Forecasts and Scenarios with the Actual Development of Renewable Energy Sources Germany-Europe-World. Agentur für Erneuerbare Energien/German Agency for Renewable Energy, Berlin May 2009.

Pimentel, D., Marklein, A., Toth, M.A., Karpoff, M.N., Paul, G.S., McCormack, R., Kyriazis, J., Krueger, T., 2009. Food versus Biofuels: Environmental and Economic Costs. Human Ecology 37, 1–12.

Raven, R.P.J.M., Gregersen, K.H., 2007. Biogas plants in Denmark: successes and setbacks. Renewable and sustainable energy reviews 11 (2007), 116–132.

REN21, Renewables, 2005. Global Status Report. Worldwatch Institute, Washington, DC 2005.

REN21, Renewables, 2010. Global Status Report. , Paris: July 2010.

REN21, Renewables, 2011. Global Status Report. REN21 Secretariat, Paris 2011.

REN21, Renewables, 2012. Global Status Report. REN21 Secretariat, Paris 2012.

REN21, Renewables Global Status Report, 2006. Update. Paris: REN21 Secretariat and Worldwatch Institute, Washington, DC 2006.

Kemp, Rene, et al., 2011. Analysis of Innovation Drivers and Barriers in Support of Better Policies: Economic and Market Intelligence on Innovation. at p87. Available at Innogrips http://www.merit.unu.edu/archive/docs/hl/201104_INNOGRIP2_Report_v29B.pdf.

Renewables 2004: www.renewables2004.de.

RE-Shaping, 2011. Review Report on Support Schemes for Renewable Electricity and Heating in Europe. Intelligent Energy Europe.

Rodriguez, J.M., Alonso, O., Duvison, M., Domingez, T., 2008. The integration of renewable energy and the system operation: The Special Regime Control Centre (CECRE) in Spain. Power and Energy Society General Meeting – Conversion and Delivery of Electrical Energy in the 21st Century, IEEE, Pittsburgh, PA, USA.

Rose, J., Webber, E., Browning, A., Chapman, S., Rose, G., Eyzaguirre, C., Keyes, J., Fox, K., Haynes, R., McAllister, K., Quinlan, M., Murchie, C., 2008. Freeing the Grid: Best and Worst Practices in State Net Metering Policies and Interconnection Standards. Network for New Energy Choices, New York, NY, USA.

SAIC, Canada., 2010. Survey of Active Solar Thermal Collectors, Industry and Markets in Canada. Science Applications International Corporation.

Sawin, J.L., 2004. National Policy Instruments: Policy Lessons for the Advancement and Diffusion of Renewable Energy Technologies Around the World—Thematic Background Paper. International Conference for Renewable Energies. Secretariat of the International Conference for Renewable Energies, Bonn, Germany (2004).

Sawin, J.L., Moomaw, W.R., 2009. Renewable Revolution: Low-Carbon Energy by 2030. Worldwatch Institute, Washington, DC.

Sawin, J.L., 2001. The Role of Government in the Development and Diffusion of Renewable Energy Technologies: Wind Power in the United States, California, Denmark and Germany, 1970-2000. Ph.D. Thesis. Fletcher School of Law and Diplomacy, Tufts University, Medford, MA, USA 672 pp.

Sawin, Janet L, 2004. Mainstreaming Renewable Energy in the 21st Century. Worldwatch Paper 169, Washington, DC May 2004.

Schwabe, Cory, Karlynn, Newcomb, James, 2009. Renewable energy project financing: impacts of the financial crisis and federal legislation. Technical Report NREL/TP-6A2–44930 Golden, CO: National Renewable Energy Laboratory, July 2009.

Searchinger, T., Heimlich, R., Houghton, R.A., 2008. Use of U.S. croplands for biofuels increases greenhouse gases through emissions from land use change. Science 319, 1238–1240.

Sematech Collaborative Model Could Sharpen Other Industries, Strategist Says. May 4, 2006. SEMATECH News, Austin TX. http://www.sematech.org/corporate/news/releases/20060504b.htm.

Sematech Facing Cut. The New York Times. February 20, 1998. Available at http://www.nytimes.com/1988/02/20/business/SEMATECH-facing-cut.html?src=pm.

Seyboth, K., Beurskens, L., Langniss, O., Sims, R.E.H., 2008. Recognising the potential for renewable energy heating and cooling. Energy Policy 36 (7), 2460–2463.

Sherwood, L., 2010. U.S. Solar Market Trends 2009. Interstate Renewable Energy Council.

Sims, R., Mercado, P., Krewitt, W., Bhuyan, G., Flynn, D., Holttinen, H., et al., 2011. Integration of Renewable Energy into Present and Future Energy Systems. In: Edenhofer, O., Pichs-Madruga, R., Sokona, Y., Seyboth, K., Matschoss, P., Kadner, S. (Eds.), IPCC Special Report on Renewable Energy Sources and Climate Change Mitigation, Cambridge University Press, Cambridge.

Sioshansi, R., Short, W., 2009. Evaluating the impacts of real-time pricing on the usage of wind generation. IEEE Transactions on Power Systems 24 (2), 516–524.

South Korea Knowledge Ministry Sets out Green Growth Strategies, http://www.korea.net/detail.do?guid=24603 (December 10, 2008).

Stephen Lacey, 2010. What Free Market? Available at http://thinkprogress.org/romm/2011/09/26/328612/new-report-energy-subsidies/2010.

Stern, N., 2007. The Economics of Climate Change. Cambridge University Press, 712pp. http://www.hmtreasury.gov.uk/sternreview_index.htm. Available at webarchive.nationalarchives.gov.uk.

Sustainable Energy Authority of Ireland: http://www.seai.ie/Your_Business/Accelerated_Capital_Allowance/.

Swedish Energy Agency, 2010. Energy in Sweden– facts and figures 2010.

Teckenburg, E., Rathmann, M., Winkel, T., Ragwitz, M., Steinhilber, S., Resch, G., Panzer, C., Busch, S., Konstantinaviciute, I., et al., 2011. Renewable energy policy country profiles. http://www.reshaping-res-policy.eu/downloads/RE-SHAPING_Renewable-Energy-Policy-Country-profiles-2011_FINAL_1.pdf.

The World Bank, 2010. Report on Barriers for Solar Power Development in India. South Asia Energy Unit Sustainable Development Department.

The World Bank: Green Bond Fact Sheet, available at http://treasury.worldbank.org/cmd/pdf/WorldBank_GreenBondFactsheet.pdf.

Thornley, Patricia, Cooper, Deborah, 2008. The Effectiveness of Policy Instruments in Promoting Bioenergy. Biomass and Bioenergy 32, 903–913.

U.S. Department of Commerce, http://www.commerce.gov/blog/2010/12/07-/renewable-energy-and-energy-efficiency-export-initiative-announced-today (December 7, 2010).

U.S. Department of Energy, Loan Programs Office, https://lpo.energy.gov/?page_id=45.

U.S. Department of Energy, Loan Programs Office, https://lpo.energy.gov/?p=4678.

U.S. Department of Energy, Loan Programs Office, https://lpo.energy.gov/wp-content/uploads/2010/09/2009-CPLX-TRANS-sol.pdf.

U.S. Department of Energy, Loan Programs Office, https://lpo.energy.gov/?projects=abo und-solar.

UNEP, 2008. Public Finance Mechanisms to Mobilise.

UNEP, Investment in Climate Change Mitigation: An overview of mechanisms being used today to help scale up the climate mitigation markets, with a particular focus on the clean energy sector. United Nations Environment Programme, Paris, France. Available at: www.unep.fr/energy/finance/documents/pdf/UNEP_PFM%20_Advance_Draft.pdf.

UNEP and BNEF, 2010. Global Trends in Sustainable Energy Investment 2010: Analysis of Trends and Issues in the Financing of Renewable Energy and Energy Efficiency. United Nations Environment Programme (UNEP). France and Bloomberg New Energy Finance (BNEF), Paris.

UNEP/BNEF, 2011. Global Trends in Renewable Energy Investments. , Paris.

UNEP/Frankfurt School/BNEF. 2012. Global Trends in Renewable Energy Investment 2012. Frankfurt School, Frankfurt.

Union of Concerned Scientists, 2011. Production Tax Credit for Renewable Energy. available at http://www.ucsusa.org/clean_energy/solutions/big_picture_solutions/production-tax-credit-for.html.

Van der Linden, N.H., Uyterlinde, M.A.,Vrolijk, C., Nilsson, L.J., Khan, J.,Astrand, K., Erics-
son, K.,Wiser, R., 2005. Review of International Experience with Renewable Energy
Obligation Support Mechanisms. ECN-C-05–025. Energy Research Centre of the
Netherlands. Petten,The Netherlands.

Verbruggen, A., Lauber,V., 2009. Basic concepts for designing renewable electricity support
aiming at a full-scale transition by 2050. Energy Policy 37 (12), 5732–5743.

Vitousek, P.M., Mooney, H.A., Lubchenco, J., Melillo, J.M. 1997. Human domination of
Earth's ecosystems. Science 277 (5325), 494–499.

Weiss, W., Mauthner, F., 2010. Solar Heat Worldwide—Markets and Contribution to the
Energy Supply 2008. IEA SHC.

Wiesenthal,T., Leduc, G., Christidis, P., Schade, B., Pelkmans, L., Govaerts, L., Georgopoulos, P.,
2009. Biofuel support policies in Europe: Lessons learnt for the long way ahead. Renewable and Sustainable Energy Reviews 13, 789–800.

Wiser, R., Pickle, S., 1997. Financing Investments in Renewable Energy: The Role of Policy Design and Restructuring. LBNL-39826. Lawrence Berkeley National Laboratory,
Berkeley, CA, USA.

Wiser, R., Barbose, G., Holt, E., 2010. Supporting Solar Power in Renewables Portfolio
Standards: Experience from the United States. Lawrence Berkeley National Laboratory,
Berkeley, CA, USA.

Wiser, R., Porter, K., Grace, R., 2005. Evaluating experience with renewables portfolio
standards in the United States. Mitigation and Adaptation Strategies for Global Change
10, 237–263.

Wiser, R., Bolinger, M., Barbose, G., 2007. Using the Federal Production Tax Credit to build a
durable market for wind power in the United States. The Electricity Journal 20 (9), 77–88.

Wong, J.L., 2010. China's Clean Energy Push - Evaluating the Implications for American
Competitiveness. available at, www.americanprogress.org/issues/2010/06/china_clean_
energy_push.html.

World Bank: http://climatechange.worldbank.org/news/renewable-energy-energy-efficiency-
financing-world-bank-group-hits-all-time-high.

World Future Council, FITness Testing, Exploring the Myths and Misconceptions about
Feed-in Tariffs, available at: http://www.futurepolicy.org/fileadmin/user_upload/PDF
/FITness_Testing_Myths.pdf.

Yin, H., Powers, N., 2010. Do state renewable portfolio standards promote in-state renewable
generation? Energy Policy 38 (2010), 1140–1149.

Yu, J., Ji, F., Zhang, L., Chen,Y., 2009. An over painted oriental arts: Evaluation of the development of Chinese renewable energy market using wind power market as model.
Energy Policy 37, 5221–5225.

REFERENCES FOR CASE STUDIES (IN ADDITION TO SOURCES LISTED ELSEWHERE)

United States (Case Study 1, 10, 18, 19)

AWEA, 2010. Windpower Outlook 2010. American Wind Energy Association (AWEA),
Washington, DC, USA.

AWEA, 2011. Fourth Quarter 2010 Market Report. American Wind Energy Association
(AWEA),Washington, DC, USA.

ERCOT, 2010. Texas Posts Record Increase in Voluntary Renewable Energy Credits:
State Exceeds Legislature's 2025 Goal 15 Years Early. Press release. Electric Reliability
Council of Texas (ERCOT), Austin,TX, USA. Available at www.ercot.com/news/press_
releases/2010/nr-05-14-102010.

Fischlein, M., Larson, J., Hall, D., Chaudhry, R., Peterson, T.R., Stephens, J., Wilson, E., 2010. Policy stakeholders and deployment of wind power in the sub-national context: A comparison of four U.S. states. Energy Policy 38 (8), 4429–4439.

Gray, Tom, 2011. The AWEA Blog: Into the Wind Stetsons off to Gov. Renewable Energy World.com, 25 August 2011 Perry on wind powerhttp://www.renewableenergyworld.com/rea/blog/post/2011/08/stetsons-off-to-gov-perry-on-wind-power?cmpid=WNL-Friday-August26-2011.

Gülen, G., Michot Foss, M., Makaryan, R., Volkov, D., 2009. RPS in Texas—Lessons Learned and Way Forward, " Center for Energy Economics, Bureau of Economic Geology. 2009 University of Texas at Austin, http://www.usaee.org/usaee2009/submissions/OnlinePr oceedings/Gulen%20et%20al.pdf2009.

REN21 Renewable Energy Policy Network, Renewables, 2005. Global Status Report. Worldwatch Institute, Washington, DC 2005.

REN21, Renewables Global Status Report, 2006. Update. Paris: REN21 Secretariat and Worldwatch Institute, Washington, DC 2006.

Wattles, Paul, 2011. ERCOT Demand Response. Powerpoint presentation for Long-Term Study Task Force, 3 May 2011, http://www.ercot.com/content/meetings/lts/keydocs/2011/0503/LTSTF%20DR%20Slides%20050311_FINAL.pdf.

Wiser, R., Bolinger, M., 2010. Wind Technologies Market Report. LBNL-3716E. U.S. 2009 Department of Energy, Washington, DC, USA. Available at eetd.lbl.gov/ea/emp/reports/lbnl-3716e.pdf.

Wiser, R., Barbose, G., 2008. Renewables Portfolio Standards in the United States: A Status Report with Data Through 2007. LBNL-154E. Lawrence Berkeley National Laboratory, Berkeley, CA, USA.

Wiser, R., Barbose, G., Holt, E., 2010. Supporting Solar Power in Renewables Portfolio Standards: Experience from the United States. Lawrence Berkeley National Laboratory, Berkeley, CA, USA.

China (Case study 2)

Han, J., Mol, A., Lu, Y., Zhang, L., 2009. Onshore wind power development in China: Challenges behind a successful story. Energy Policy 37 (8), 2941–2951.

Ku, J., Baring-Gould, E.I., Stroup, K., 2005. Renewable Energy Applications for Rural Development in China. NREL/CP-710–37605. National Renewable Energy Laboratory, Golden, CO, USA 5 pp.

Martinot, E., Junfeng, Li., 2007. Powering China's Development – The Role of Renewable Energy. Worldwatch Institute, Washington, DC, USA.

Martinot, E., 2010. Renewable power for China: Past, present, and future. No. 3 Frontiers of Energy and Power Engineering in China Vol. 4, 287–294.

REN21, 2009. Recommendations for Improving Effectiveness of Renewable Energy Policies in China. Renewable Energy Policy Network for the 21st Century (REN21). France, Paris.

U.S. National Renewable Energy Laboratory (NREL), 2004. Renewable Energy Policy in China: Financial Incentives. NREL/FS-710-36045, Golden, CO April 2004.

Wallace, W.L., Jingming, L., Shangbin, G., 1998. The Use of Photovoltaics for Rural Electrification in Northwestern China. NREL/CP-520–23920. National Renewable Energy Laboratory and Chinese Ministry of Agriculture, Golden, CO, USA and Beijing, China.

Wang, F., Haitao Yin, S., 2010. China's renewable energy policy: Commitments and challenges. Energy Policy 38, 1872–1878.

Yu, J., Ji, F., Zhang, L., Chen, Y., 2009. An over painted oriental arts: Evaluation of the development of Chinese renewable energy market using wind power market as model. Energy Policy 37, 5221–5225.

Brazil (Case Study 3, 7)

Azevedo, J.M., Galiana, F.D., 2009. The sugarcane ethanol power industry in Brazil: Obstacles, success and perspectives. In: IEEE Electrical Power & Energy Conference, Canada, Montreal.

Cerri, C.E.P., Easter, M., Paustian, K., Killian, K., Coleman, K., Bemoux, M., Falloon, P., Powlson, D.S., Batjes, N.H., E.Milne, Cerri, C.C., 2007. Predicted soil organic carbon stocks and changes in the Brazilian Amazon between 2000 and 2030. Agriculture. Ecosystems and Environment 122, 58–72.

Dias de Moraes, M.A.F., Rodrigues, L., 2006. Brazil Alcohol National Program. Relatório de pesquisa. Piracicaba, Brazil 54.

Goldemberg, J., 2006. The ethanol program in Brazil. Environmental Research Letters, 1, 014008.

Goldemberg, J., 2009. The Brazilian experience with Biofuels. Innovations 4 (4), 91–107.

Goldemberg, J., Coelho, S.T., Nastari, P.M., Lucon, O., 2004. Ethanol learning curve—the Brazilian experience. Biomass and Bioenergy 26 (3), 301–304.

Mathews, John A., Goldsztein, Hugo, 2009. Capturing latecomer advantages in the adoption of biofuels: The case of Argentina. Energy Policy 37, 326–337.

Riegelhaupt, Enrique with Paloma Manzanares, Mercedes Ballesteros, Coelho, Suani, Guardabassi, Patricia, James, Carlos Saint, Aroca, Germán, Rutz, Dominik, Janssen, Rainer, 2009. Overview of biofuel markets and biofuel applications in Latin America. prepared for the BioTop project, supported by the European Commission in the 7th Framework Program, Mexico April 2009.

Walter, A., 2006. Is Brazilian Biofuels Experience a Model for Other Developing Countries? Entwicklung & Ländlicher Raum 40, 22–24.

Thailand (Case Study 4)

Amranand, P., 2008. Alternative Energy, Cogeneration and Distributed Generation: Crucial Strategy for Sustainability of Thailand's Energy Sector. Energy Policy and Planning Office (EPPO), Ministry of Energy, Bangkok, Thailand.

Amranand, P., 2009. The role of renewable energy, cogeneration and distributed generation in sustainable energy development in Thailand. Keynote Address World Renewable Energy Congress 2009 Asia, BITEC, Bangkok, Thailand.

Fox, J., 2010. Renewable Energy in Thailand: Green Policies Take Off. Thailand Law Forum. Available at , www.thailawforum.com/green-policies-take-off.html.

Greacen, C., Greacen, C., Plevin, R., 2003. Thai power: Net metering comes to Thailand. ReFocus 4 (6), 34–37. Available at netmeter.org/en/docs/NetMeteringRefocusNov20 03.pdf.

Yoohoon, A., 2010. Low Carbon Development Path for Asia and the Pacific: Challenges and Opportunities to the Energy Sector. ESCAP Energy Resources Development Series ST/ESCAP/2589, Beijing, China.

Spain (Case Study 5, 11)

Renewable Energy Magazine, 2011. Solar industry takes to streets to protest against retroactive tariff cuts in Spain, January 2011. http://www.renewableenergymagazine.com/rene wableenergy/magazine/index/pag/blog/botid/1/colright/blog/tip/articulo/tag/Fotov oltaica/pagid/13077/title/Solar%20industry%20takes%20to%20streets%20to%20protes t%20against%20retroactive%20tariff%20cuts%20in%20Spain/#slide_8.

Bravo, Isidoro Lillo, 2004. Procedimiento de conexion para inyectar energia a la red electrica convencional desde una instalacion fotovoltaica, " Grupo de Termodinámica y Energías Renovables, Escuela Superior de Ingenieros. undated. (cited in Sawin University of Seville, Spain.

Fulton, Mark, Bruce M., Kahn, Nils Mellquist, Emily Soong, Jake Baker, Lucy Cotter, 2009. Paying for Renewable Energy:TLC at the Right Price.Achieving Scale Through Efficient Policy Design. DBCCA , Deutsche Bank Climate Change Advisors, December 2009.

IDAE, 2008. Seguimiento del Plan de Energías Renovables en España (PER) 2005-2010. Memoria 2008. Institute for the Saving and Diversification of Energy (IDAE). Ministerio de Industria,Turismo y Comercio, Madrid, Spain.

IDAE, 2010. La industria fotovoltaica española en el contexto europeo. Institute for the Saving and Diversification of Energy (IDAE). Ministerio de Industria,Turismo y Comercio, Madrid, Spain.

MITyC, 2008. Real Decreto 1578/2008, de 26 de septiembre, de retribución de la actividad de producción de energía eléctrica mediante tecnología solar fotovoltaica para instalaciones posteriores a la fecha límite de mantenimiento de la retribución del Real Decreto 661/2007, de 25 de mayo, para dicha tecnología. (in Spanish). Boletín Oficial del Estado, Madrid, Spain. Available at www.boe.es/aeboe/consultas/bases_datos/doc.php?id=BOE-A-2008-155952008.

Oettinger, G., Hedegaard, C., February 2011. Letter to D. Miguel Sebastián, Spanish Minister of Industry. (EU Commissioners) Tourism and Commerce, Madrid, dated 22. http://ec.europa.eu/commission_2010-2014/hedegaard/headlines/docs/letter_sebastian_es.pdf.

Solas Power, 2011. Fiscal deficit forces Spain to slash renewable energy subsidies. 22 July 2011, http://solaspower.com/page2/page2.php?categories=Spain2011.

Stuart, B., 2011. PV Legal: Positive PV progress made in legal and administrative barriers. PV Magazine. http://www.pv-magazine.com/news/details_/beitrag/pv-legal–positive-pv-progress-made-in-legal-and-administrative-barriers_100002956/8/.

Witowski, P., 2011. Spain's solar industry protests against retroactive subsidy cuts. Platts. http://www.platts.com/RSSFeedDetailedNews/RSSFeed/ElectricPower/8452448.

EU (Case Study 6, 9, and 17)

Euractiv: Doubts cloud launch of EU biofuels sustainability schemes. http://www.euractiv.com/climate-environment/doubts-cloud-launch-eu-biofuels-sustainability-schemes-news-506702.

European Commission, Jan 2011. Commission Staff Working Document; Recent progress in developing renewable energy sources and technical evaluation of the use of biofuels and other renewable fuels intransport in accordance with Article 3 of Directive 2001/77/EC and Article 4(2) of Directive 2003/30/EC.Accompanying document to the Communication from the Commission to the European Parliament and the Council: Renewable Energy: Progressing towards the 2020 targethttp://ec.europa.eu/energy/renewables/reports/doc/sec_2011_0130.pdf.

Sweden (Case study 8)

Bergek, A., Jacobsson, S., 2010. Are tradable green certificates a cost-efficient policy driving technical change or a rent-generating machine? Lessons from Sweden 2003–2008. Energy Policy 38 (3), 1227–1606.

Ericsson, K., Svenningsson, P., 2009. Introduction and Development of the Swedish District Heating Systems: Critical Factors and Lessons Learned. Lund University, Lund, Sweden.

IEA-RETD, 2008. Renewables for Heating and Cooling—Untapped Potential.

International Energy Agency, 2009. Cogeneration and District Energy: Sustainable Energy Technologies for Today and Tomorrow. International Energy Agency and Organisation for Economic Co-operation and Development, Paris, France 24 pp.

REN21, Renewables, 2010. Global Status Report. , (Paris: REN21, 2010).

Swedish Energy Association (SEA), 2009. F. acts and figures—Energy in Sweden 2009 ET2009: 29. Swedish Energy Agency, Eskilstuna.

Thornley, Patricia, Cooper, Deborah, 2008. The Effectiveness of Policy Instruments in Promoting Bioenergy. Biomass and Bioenergy 32, 903–913.

Germany (Case Study 12)

BMU, 2006. German Federal Ministry for the Environment, Nature Conservation and Nuclear Safety (BMU), Graphics and tables on the development of renewable energy sources in Germany in 2005. May 2006, http://www.erneuerbare-energien.de/files/english/renewable_energy/downloads/application/pdf/ee_frhjahr_2006_eng.pdf.

BMU, 2011. German Federal Ministry for the Environment, Nature Conservation and Nuclear Safety (BMU), The path to the energy of the future—reliable, affordable and environmentallysound. June 2011, http://www.bmu.de/english/energy_efficiency/doc/47609.php2011.

BMU, 2011. German Federal Ministry for the Environment, Nature Conservation and Nuclear Safety (BMU). July 2011. Development of Renewable Energy Sources in Germany in 2010, http://www.bmu.de/files/english/pdf/application/pdf/ee_in_deutschland_graf_tab_en.pdf.

BMU, 2011. German Federal Ministry for the Environment, Nature Conservation and Nuclear Safety (BMU). Federal Environment Minister Röttgen: 20 percent renewable energies are a great success, Berlinhttp://www.erneuerbare-energien.de/inhalt/47744/3860/ press release no. 108/1130 August 2011.

Büsgen, Uwe, Wolfhard Dürrschmidt, 2009. The expansion of electricity generation from renewable energies in Germany. A review based on the Renewable Energy Sources Act Progress Report 2007 and the new German feed-in legislation. Energy Policy 37, 2536–2545.

Fulton, Mark, Nils Mellquist, 2011. The German Feed-in Tariff for PV: Managing Volume Success with Price Response. Deutsche Bank Group 23 May 2011.

Judith, Lipp 2007. Lessons for effective renewable electricity policy from Denmark, Germany and the United Kindom. Energy Policy 35, 5481–5495.

Poeschl, Martina, Shane, Ward Philip, Owende, 2010. Prospects for expanded utilization of biogas in Germany. Renewable and Sustainable Energy Reviews 14, 1782–1797.

Denmark (Case Study 13)

BTM Consult ApS, 2010. World Market Update 2009. BTM Consult ApS, Ringkøbing, Denmark.

Danish Energy Agency, Danish Energy Policy 1970-2010. Vision: 100% Independence of Fossil Fuels, Copenhagen (undated), http://www.ens.dk/en-US/Info/news/Factsheet/Documents/DKEpol.pdf%20engelsk%20til%20web.pdf.

Danish Ministry of the Environment, 1993. Cirkulære om primærkommuners planlægning for vindmøller (til alle kommunalbestyrelser). Miljøiministeriet, Copenhagen, Denmark.

Danish Wind Industry Association, 2010. Danish Wind Industry Annual Statistics 2010. Frederiksberg (undated), http://www.e-pages.dk/windpower/15.

Danish Wind Industry Association, 2010. Danish Wind Industry Maintains High Export Figures In 2009 Despite Financial Crisis. Danish Wind Industry Association (Vindmølleindustrien) Available at www.windpower.org/en/news/news.html, Frederiksberg, Denmark.

ECOFYS, Fraunhofer, Energy Economics Group, and Lithuanian Energy Institute, 2011. Renewable Energy Policy Country Profiles, 2011 version, prepared within the Intelligent Energy Europe project (RE-Shaping).

Judith, Lipp 2007. Lessons for effective renewable electricity policy from Denmark, Germany and the United Kingdom. Energy Policy 35, 5481–5495.

Madsen, B.T., 2009. Public initiatives and industrial development after 1979. The Danish Way: From Poul La Cour to Modern Wind Turbines, Poul La Cour Foundation, Askov, Denmark.

Richardson, Katherine, Dahl-Jensen, Dorthe, Elmeskov, Jørgen, Hagem, Cathrine, Henningsen, Jørgen, Korstgård, John, Kristensen, Niels Buus, Morthorst, Poul Erik, Olesen, Jørgen E., Wier, Mette, Nielsen, Marianne, Karlsson, Kenneth, 2011. Denmark's Road Map for Fossil Fuel Independence. The Solutions Journal Vol. 2 Issue 4, July 2011.

Upper Austria (Case Study 14)

O.O. Energiesparverband, undated. "1,000,000 m2 of installed solar collectors in Upper Austria, " www.oec.at/en/energy-in-upper-austria/solar-energy/.

O.O. Energiesparverband, undated. Biomass in Upper Austria. www.oec.at/en/energy-in-upper-austria/wood-pellets-biomass/.

Egger, C., Öhlinger, C., undated. 25% heat from solar thermal in 2030 – Upper Austria's Solar programme, O.O. Energiesparverband, Linz, Austria.

Christiane Egger and Christine Öhlinger (undated), Electricity Efficiency Policy in Upper Austria, O.O. Energiesparverband, Linz, Austria.

Egger, C., Auinger, B., Öhlinger, C., 2009. Regional Report: The RES-H/C Market in Upper Austria. A report prepared as part of the IEE project "Policy development for improving RES-H/C penetration in European Member States (RES-H Policy). O.O. Energiesparverband, Linz, Austria June 2009.

Egger, C., Ohlinger, C. Auinger, B., Brandstatter, B., Dell, G., undated, Carrots, Sticks and Tambourines: How Upper Austria Became the World's Leading Solar Thermal Marke.

IEA-RETD, 2010. Best Practices in the Deployment of Renewable Energy for Heating and Cooling in the Residential Sector.

IEA-RETD, 2010. Renewable Energy for Residential Heating and Cooling – Policy Handbook.

Güssing (Case Study 15)

Energie+, magazine: Güssing energiueautonoom. June 2011.

Guessing, Austria, World Future Council, www.futurepolicy.org/2829.html (accessed 5 March 2012).

Hurriyet Daily News (Turkey), Austrian town exports energy model, 7 May 2009, http://www.hurriyetdailynews.com/default.aspx?pageid=438&n=austrian-town-exports-energy-model-2009-07-05.

Janet L. Sawin, Moomaw, William R., 2009. An Enduring Energy Future, " in State of the World 2009: Into a Warming World. Linda Starke, W.W. Norton & Company, New York.

Municipality of Güssing: http://www.guessing.co.at/ (accessed 6 March 2012).

Rizhao (Case Study 16)

http://en.wikipedia.org/wiki/Rizhao (accessed Jan 2012).

Scientific American: Sunrise on China's First Carbon-Neutral City. (by David Biello, August 2008) http://www.scientificamerican.com/article.cfm?id=sunrise-on-chinas-first-carbo-neutral-city.

UNEP: about Rizhao. http://www.unep.org/climateneutral/Participants/Cities/Rizhao/tabid/205/Default.aspx.

Worldwatch Institute, "Solar Powered City, " in State of the World 2007, ed. Linda Starke (New York: W.W. Norton and Company, 2007). www.worldwatch. org/node/4752.

Republic of Korea (Case Study 20)

Asia Economic Institute, 2010. South Korea's Green Deal. www.asiaecon.org.

Han, P.S., 2010. Sustainable and green tourism. http://www.oecd.org/cfe/tourism/45558102.pdf.

Huffington, Post, 2009. Green New Deal For South Korea: $38.1 Billion.

UNEP, 2009. Japan and the Republic of Korea Launch Green New Deals. http://www.unep.org.

EU (Case Study 21)

European Investment Fund: Competitiveness and Innovation Framework Programme (CIP) - EU Guarantees. http://www.eif.org

European Commission, 2007. Notice of implementation of the High Growth and Innovative SME Facility under the Competitiveness and Innovation Framework Programme. (2007-2013). (2007/C 302/09).

UK (Case Study 22)

UK Department for Business Innovation & Skills, 2011. Vision for Worlds first dedicated Green Investment Bank published. http://www.bis.gov.uk/greeninvestmentbank.

UK Department for Business, Innovation & Skills, 2011. Update on the Design of the Green Investment Bank. http://www.bis.gov.uk/assets/biscore/business-sectors/docs/u/11-917-update-design-green-investment-bank.pdf.

Relevant RETD Studies

IEA-RETD (2006), Renewable Energy for Dummies— Why you Don't Use Renewable Energy and Why you Should!

IEA-RETD (2008), Innovative Electricity Markets to Incorporate Variable Production

IEA-RETD (2008), RE-EDUCATION A Scoping Study into Renewables and Education

IEA-RETD (2008), Renewable Energy Costs and Benefits for Society (RECABS)

IEA-RETD (2008), The True Costs and Benefits of Renewable Energy— Make Optimal Investment Decisions Based on Better Knowledge About the Long Term Benefits of Renewables

IEA-RETD (2009), Better Use of Biomass for Energy

IEA-RETD (2010), Renewables in Transport (RETRANS)

IEA-RETD (2010), RETRANS—Opportunities for the Use of Renewable Energy in Road Transport

IEA-RETD (2011), Achieving Climate Stabilization in an Insecure World: Does Renewable Energy Hold the Key?

IEA-RETD (2011), Climate Change Adaptation, Damages and Fossil Fuel Dependence—An RETD Position Paper on the Costs of Inaction

IEA-RETD (2011), Opportunities for the Use of Renewable Energy in Road Transport in North America, Europe and China (RETRANS2)

IEA-RETD (2011), Risk Quantification and Risk Management in Renewable Energy Projects (RISK)

IEA-RETD (2011), Strategies to Finance Large-Scale Deployment of Renewable Energy Projects: an Economic Development and Infrastructure Approach (FINANCE-RE)

IEA-RETD (2011), Sustainable Energy Scenarios and the Role of Renewable Energy

IEA-RETD (2011), Accelerating the Deployment of Offshore Renewable Energy Technologies (ADORET; published by Earthscan)

IEA-RETD (2011), Synergies between Renewable Energy and Fresh Water Production (RENWA)

IEA-RETD (2012), Business Models for Renewable Energy in the Built Environment

IEA-RETD (2012), Renewable Energies for Remote Areas and Islands (REMOTE)

Abbreviations and Acronyms

ACES	Achieving Climate and Energy Security Scenario from IEA-RETD
AWEA	American Wind Energy Association
BNEF	Bloomberg New Energy Finance
CAPEX	Capital Expenditures
CCS	Carbon Capture and Storage
CDM	Clean Development Mechanism
CEFIA	Clean Energy Finance and Investment Authority, Connecticut
CHP	Combined Heat and Power
CSP	Concentrating Solar Thermal Power
CTE	Spain's Technical Building Code (Código Técnico de la Edificación)
DECC	UK's Department of Energy and Climate Change
DH	District Heating
DOE	US Department of Energy
EC	European Commission
EEG	Germany's Renewable Energy Sources Act
EIB	European Investment Bank
EIF	European Investment Fund
EPIA	European Photovoltaic Industry Association
EREC	European Renewable Energy Council
ET-RAM	Emerging Technology Renewable Auction Mechanism
ETS	Emission Trading Scheme
ETSAP	IEA's Implementing Agreement on Energy Technology Systems Analysis Programme
EU	European Union
EV	Electric Vehicle
FIPP	U.S. Financial Institution Partnership Program
FIT	feed-in tariff
G8	Group of Eight – forum for the governments of the eight largest economies
GDP	Gross Domestic Product
GHG	Greenhouse gas

GIF   European High Growth and Innovative small-medium
                  enterprise Facility
GWEC   Global Wind Energy Council
H/C   Heating and cooling
ICLEI   ICLEI - Local Governments for Sustainability
IEA   International Energy Agency
IEA-RETD   International Energy Agency's Implementing Agreement
                  on Renewable Energy Technology Deployment
IPCC   Intergovernmental Panel on Climate Change
ISEP   Institute for Sustainable Energy Policies
ITC   Investment Tax Credit
JI   Joint Implementation
LCA   Life Cycle Analysis
LCOE   Levelized Cost of Energy
MAP   Germany's Market Incentive Program Marktanreizpro-
                  gramm
NFFO   UK's Non-Fossil Fuel Obligation
NGO   Non-governmental organizations
NIMBY   Not in my backyard
NPIC   India's National Program on Improved Chulhas
NREAP   National Renewable Energy Action Plans as part of the EU
                  Renewable Energy Directive
OECD   Organisation for Economic Co-operation and Develop-
                  ment (of 34 countries)
PACE   Berkeley's (California) Property Assessed Clean Energy
PTC   U.S. Production Tax Credit
PV   Photovoltaics OR Photovoltaic solar energy
R&D   Research and development
REC   Renewable Energy Credit
REDD   Reducing Emissions from Deforestation and Forest Degra-
                  dation
REN21   renewable energy network
RETD   renewable energy technology development (IEA Imple-
                  menting Agreement)
RHI   UK's Renewable Heat Incentive
RPS   Renewable Portfolio Standard
SDC   Small Distributed Capacity
SRREN   the IPCC's Special Report on Renewable Energy Sources
                  and Climate Change Mitigation

TREC	(Swedish) Tradable Renewable Electricity Certificates
UAE	United Arab Emirates
UN	United Nations
UNEP	United Nations Environment Programme
USD	US dollar
WBG	World Bank Group
WEO	World Energy Outlook
WTO	World Trade Organisation
WWF	World Wildlife Fund
ZEV	Zero Emission Vehicle

Units

J	Joule
toe	tonne of oil equivalent
ppm	parts per million

Prefixes

M	mega 10^6
G	Giga 10^9
T	Tera 10^{12}
P	Peta 10^{15}
E	Exa 10^{18}

SUBJECT INDEX

Note: Page numbers followed by "f", "t" and "c" indicate figures, tables and case studies respectively

A

Accelerated deployment, policy actions for, xiii–xv
  changes, need for, xvi–xvii
  current trends, xvii–xviii
  energy investment, new priorities, xiii–xix
  policy makers, role in, xivc
Accelerated depreciation, 75
  India's experience in, 84–85
ACES. *See* "Achieving Climate and Energy Security"
ACES scenario
  electricity sector
    in, 41–42
  IEA-RETD, 40–41
  overview, 32t–33t
"Achieving Climate and Energy Security" (ACES), 29
ACTION Star, iii–xiv, iiic
ACTION Star
  recommendations, xxv
  overarching guidelines, xxv–xxvi
Adverse and unsustainable impacts, addressing of, 56–57
Alliance building, to lead paradigm change, xxvi
American Wind Energy Association (AWEA), 61, 76c–78c
Arab awakening, 56
"Arab Spring", xiv–xv, 56
Asia-Pacific Partnership on Clean Development and Climate, 52t–53t

B

"Bio-based economy", 197–198
Bio-ethanol, 207–208
Biodiesel, 11, 207

Bioenergy, 205–206
  large-scale bioenergy applications, 206
  medium-scale bioenergy applications, 206
  small-scale bioenergy applications, 206
Biofuel(s), 103
  biodiesel, 207
  EU Directive, 108c
  Mandates, 107–110
  national/international policy, 104
  policies. *See* Renewable transportation
  sustainability, 103, 110
Biogas, use of, 8
Biomass, use of, 195–196
  logistics, 196
Biomass conversion routes, 197f
Biomass energy, 7–8
  combined heat and power (CHP), 8
  global biomass power capacity, 8
  and sustainability, 57–60
Biomass energy system, 195
  logistics, 196
Biomass heat, 11–12
  pellets, 12
  possible conversion routes for, 207f
Biomass production, 195
Biomass sources, 196f
Bloomberg New Energy Finance (BNEF), 13–14
  renewable energy investments, xix
  wind power, global investment, xvii
Blue Map scenario, 39
  Organization for Economic Cooperation and Development (OECD) countries versus non-OECD countries, 37–38
Brazil's Renewable Energy Tender, 90c
Brazil's Modern Ethanol Program, 109c–110c